The Language of Design

Andy Dong

The Language of Design

Theory and Computation

 Springer

Andy Dong, PhD
Design Lab
Faculty of Architecture, Design and Planning
University of Sydney
Wilkinson Building
148 City Road
Darlington
Sydney
NSW 2006
Australia

ISBN 978-1-84996-816-4   e-ISBN 978-1-84882-021-0

DOI 10.1007/978-1-84882-021-0

A catalogue record for this book is available from the British Library

Cover design: eStudio Calamar S.L., Girona, Spain

Printed on acid-free paper

9 8 7 6 5 4 3 2 1

springer.com

Preface

The imperative of design is being galvanized by the increasing consumerization of 'good' design. The packaging for my Apple products reminds me that they were "Designed by Apple in Cupertino, Calif." Advertisements for David Jones, an upscale department store in Australia, remind me that David Jones support Australian designers (notwithstanding a few New Zealand-born designers re-territorialized as Australian). Companies such as Ikea and Target market affordable 'good' design to middle-class consumers in developed countries and in developed regions of developing countries. 'Masstige' brands such as Alessi, G-Star, and Puma propel the status of designers such as Michael Graves, Marc Newsom, and Philippe Starck to near pop-culture icons. Magazines focusing on good design such as *Wallpaper* create a global state of mind about design. Design matters, but which designs matter and how design matters is imaged by what people read about design and what is being said about design.

All of this dialogue about 'good' design is fairly removed from those whose well-being depends on their capability to design rather than being on the receiving end (or, more to the point, purchasing end) of 'good' design. From squatters as slum-developers in Mumbai to the Solidarity for the Urban Poor Federation in Cambodia, the urban poor are involved in ambitious programs to improve their well-being by taking part in the design of housing projects and sanitation schemes. For the urban poor in developing countries and regions, design matters, too. It is the basis for their well-being. And we should not be remiss to forget all of the multimedia and Web designers creating digital spaces in arenas such as MySpace, YouTube, and Wikipedia. So significant were they that in 2006 the Editors of *Time Magazine* named them ("You") their "Person of the Year".

Thus, on the one hand, there are ongoing dialogues about the meaning of design, the design identity of star designers, and the practices that distinguish 'expert' designers from 'novice' designers. On the other hand, there are those who clearly engage in what could be legitimately construed as design, yet few would label a slum dweller a 'designer', not at least in the terminology and attitudes toward design portrayed in the popular and academic literature. Even if the slum dweller (or dwellers collectively) managed to achieve the design of suitable housing and sanitation where outside expertise had previously failed to do so, one would be hard-pressed to find them profiled in *I.D.* The rise of personal digital content creation has already stirred multinational media companies to include these 'anything goes' designers into their portfolio.

Part of the disconnect between these enactments of design lays in what is construed as design, that is, the definition of design, the state of being of design. Are these camps engaging in design? They might be if we could agree on a definition of design, but that seems unlikely nor is it necessarily desirable. At the INDEX[1] 2005 awards in Denmark, participants were asked to define design. Each of the definitions eloquently defined a position toward design, echoing themes about the production or expression of a work such as functionality, meaning, social relevance, transforming ideas into forms, etc., which are commonly held notions about design. But, no two of the definitions could be said to agree on what design is.

This may not be the real problem we face, though. Our energies may be wasted in trying to define design. Here, we might, as the post-structuralist thinkers have done, not dwell on the state of being of design, demanding a definition of design, but rather think about the 'becoming-design' as Deleuze and Guattari theorize in their book *A Thousand Plateaus*. For Deleuze and Guattari, the question of becoming rather than the question of being allows us to escape the trap of models that try to hierarchically organize the world. Their concept of becoming focuses on the complex relations by which objects can be conceptualized. The becoming of the 'Other' that Deleuze and Guattari write about seems right when thinking about design because design is always about transforming. And the conditions for becoming seem rather similar to the conditions for design. In becoming, there is always an interior (e.g., 'man'), an exterior (e.g., 'animal') and a "line of deterritorialization" that passes between these two "forming so many becomings between things, or so many lines of deterritorialization" (Deleuze and Guattari 1987, p. 294) that transform becoming-man through becoming-animal. In design, there is always a problem space, a solution space, and the differential and dynamic processes that transform one space into the other. Where can this notion of becoming-design take us in thinking about design? And, how might we enter into the becoming-design?

In this book, we turn to this issue of the becoming-design by examining another mode of becoming – the language of design. Designing is certainly a language on its own, partly performing what cannot be conversed or said but only enacted by designing. The designed work is also a language on its own, giving us accounts on its identity and states of being through a visual vocabulary. What is said and written about design give expression to the manner of actualization of the designed work – the becoming-design – the inputs into the activities from which the designed work was produced and can be appreciated, and how the designed work was dissolved and formed through the motions and emotions of design practice.

---

[1] The INDEX Awards sponsored by the Danish Government recognize designed works and design programs that improve the lives of people around the world.

It is the seductive and sensual pleasure of experiencing becoming-design that after all lures us to these stories about design. Printed and verbal accounts of designed works form an image of the vector of forces that enact design. This image of the designed work expressed in language makes it is possible for us to fantasize the becoming-design. I think this is what is meant by Deleuze and Guattari in thinking about becoming. Becoming-design is about perceiving and anticipating what a designed work is not yet, but will become. And even the final designed work is not static – it is always becoming. It can become an Other through mass customization by users or indefinitely cycled 'cradle-to-cradle' into other designed works as William McDonough implores.

The situation I related above is akin to what David Brooks wrote in his book *Bobos in Paradise*. It is not enough any more for us to buy an orange juicer. We want a post-modern treatise on the juicer; we want to know about the NURBS equation for the curvature of the juicer; we want to know what school of design the juicer belongs to. For me, reading design texts reminds me of touching the lightning balls I used to play with in kitschy tourist shops dotting America. These lightning balls produce streams of electrically charged particles that jump from the center to your palm or fingers as you touch the glass exterior. Each of these streams of charged particles seems like one of these vectors. With every read, as with every touch, one of the vectors of power recalls an image of thought about a designed work.

Our aim in this book is to examine the language of design in order to grapple with becoming-design. We approach this issue by problematizing how the language of design constitutes an account of designing and the designed work. For quite some time now, cognitive science and social science have served as the bases for predominant approaches to understanding design. In my own field of design computing, understanding how designers design through the methods of cognitive science has inspired models of computing, which in turn serve as the foundation for design tools and computational systems that enact design. The homology between design and cognition (models of human mental processing) are taken as a given because design (at least in its purposeful and intentional senses) seems to be a uniquely human endeavor. What is somewhat problematic about the cognitive science and social science approaches is their tendency to reduce design to repeatable steps. The idea is that if we can learn how it is that designers do design, then we can, in a sense, codify the process and thereby repeat the process reliably. Moreover, these steps can be encoded computationally so that computers could enact design. It is undeniable that these approaches have generated a wide variety of useful computational tools and systems, and I certainly do not discount the fruitfulness of these approaches. What this book questions, however, is whether knowing design by understanding actions undertaken in design thereby leading to faithful repetitions of design simply lets us know what design is without knowing what design feels like. If we want to know design, perhaps we need to become-design, not just imitate design, but instead enter into the compositioning of design.

The examination of design texts will be our mechanism for entering into the becoming-design. Computational interpretations of design texts will help us render perceptible what these texts produce – the designed work – and the transformations (i.e., multiple becomings) that occurred as a result of a set of powers operating behind the texts.

Our project is informed by the post-structuralist view that texts are an expression or representation of something else, some other extra-lingual phenomena, some other effect. While motivated and guided by post-structural philosophy in thinking about how the language of design harnesses and represents that which can be conversed and said, thereby producing the act of designing and the consequent designed work, our method of analysis will be drawn from the techniques of computational science. Our conjecture is that design partially subsists in language; the substrate is the language of design. Entering into the language of design lets us inhabit the becoming-design through the becoming of another mode – language.

Once we get past the seductions of the accounts of designing and the designed work in language form, the language of design reminds us that design is an assemblage, continually made anew each time, and reiteratively deployed through various forms of discourse about design. It is re-thinking the 'designer' and what allows the designer to perform designing in a certain way as not necessarily a 'willful' choice but one that is anchored into what is said and written about design that I hope to provoke you into considering. That is, we need to think about *becoming-design* not being design.

At this point, one should be thinking why focus on language. When one thinks of designing, language is probably not the first type of representation thought of; visual forms are if you are an architect or user interface designer, equations and diagrams if you are an engineer. Moreover, designers produce representations in various formats including drawings, equations, diagrams, and multimedia. More often than not, though, these representations are accompanied by language-based descriptions. Language is a medium by which designers give an account of design and almost always accompany visual forms. My assumption, which may be optimistic, is that the practice of design may be studied more or less independently of these other representations because language is intertwined in the ontological circuit of recognition between a designed work, the designers, and the users. How the activities, methods and practices of design work to constitute the materiality of the designed work through language and how the language itself is designing are one focus of this book. Second, language produces a common sense that anchors designers and their work to a body of knowledge and practice. The language tells stories about design, which design stories to relate, and how to tell them, and how to construe the meaning of design. Thus, analyzing design as produced by the discursive deployments of the language of design may turn our attention toward design practice and the designed work as the effect of what Judith Butler described as 'reiterative performances' and their institutions that confer authority on certain work practices as being identifiable as design and on certain objects and environments as being designed.

Studying the language of design has always been problematic because the language often refers to things which do not yet exist and which may never materialize. My hope is that by thinking of the language of design as part of and enacting design, we can register the ways in which the constitution of design relies on language beyond cataloging types of design conversations and statements as "elements of the language of designing" (Schön 1983, p. 95). Following Michel Foucault's thoughts on language from *The Order of Things*, our analysis of the language of design moves "towards the place where things and words are conjoined in their common essence, and which makes it possible to give them a name." (Foucault 1994, p. 117) Rather than mapping design into deterministic and proceduralized structures, language is seen as expressing the possibilities of design, channeling chance and patterning words across a series of journeys where language is design's structuring structure.

Using This Book

Interested readers are invited to download sample MATLAB® code from a Web page accompanying this book. The sample code implements the Computational Implementations described in the chapters.

Go to `http://www.arch.usyd.edu.au/~adong/book/lod/` to download the code.

Acknowledgements

Writing, like designing, involves multiple stakeholders. This project is no exception.

It is difficult to trace the origins of any project, but one event stands out for me. My friend and colleague Petra Gemeinböck, upon her arrival into Sydney, house sat for me while I went away on a conference. That night, after helping Petra to connect to my wireless network, my friend David McInnes and Petra engaged in a long discussion on performativity. That discussion galvanized my interest in performativity. I have benefited enormously from David and Petra, who both introduced me to and tutored me on Butlerian performativity. Cristyn Davies organized tickets for me to attend Judith Butler's lecture at Angel Place in Sydney as part of the Judith Butler Symposium at the University of Western Sydney. David also introduced me to the theories of Basil Bernstein whose ideas permeate the last chapters of this book. My friendship with him has left an inexorable intellectual mark.

Even earlier thinking on this book started when I was a postdoctoral researcher at the University of California, Berkeley, with Professor Alice Agogino. She, a Master of Science student Andrew Hill, and I started formulating ideas about using computational linguistic algorithms to address the problem of monitoring and coaching student design teams. It was Andrew Hill who first suggested of the idea of using latent semantic analysis and with whom I was able to make the first working system that could analyze design text for linguistic characteristics associated with successful creative design work. The quality of this idea was recognized by the Design Research Society and Elsevier Ltd. by awarding me the Design Studies Prize in 2005 for my publication *The latent semantic approach to studying design team communication*. Along with the prize of £500 and the recognition in my community, the award motivated me to think that the study of language and design, computationally, might just be a worthwhile pursuit.

The work on lexical chain analysis was greatly advanced by the astute and careful programming by my research assistant and friend Kevin Davies. That his code has been ported across computers and operating systems attests to the great care he took in implementing the algorithms. Justin Clayden took all of the code, a draft user interface and inscrutable Matlab code, and wrote a distributable software package that could be used by other researchers. He added in text processing code to make it accessible to, as he says, "non-geeks". Colleagues at the University of Bath, Massachusetts Institute of Technology, and Nanyang Technological University tested the software and offered informative feedback.

Xiong Wang came to me as a PhD student in 2005 with an interest in furthering his research in support vector machines. Little did he know that I would throw to him the task of using support vector machines to analyze appraisals in design text and implement my language model of appraisal. He took all of the changes to the model in stride; I am thankful for his patience in a rather quixotic young academic with an idea that, at times, did not seem stable.

Maaike Kleinsmann arrived from Delft University of Technology as a Visiting Researcher just in time to help me operationalize the language of appraisal in design and to test it on real data. She spent hours with me analyzing transcripts of designers working and clarifying the language model and rules to analyze the language. The language model for appraisal is better for her contributions, and any continued deficiencies stem from my intransigence to make further changes. Neil McCann designed the illustration which appears on the cover and in Fig. 2.2 and played language games with me while I was writing. Rachel AC Luck read an early draft of the book and gave poignant comments and invaluable moral support to the overall project. Other people, who I only know via the Internet, sent in e-mails of support with questions and suggestions on a Web-based working copy. All projects benefit from a good bout of good luck, and I am fortunate to have my share from the kindness of strangers and colleagues alike.

It would not have been possible to validate the results of the computational linguistic algorithms without the assistance of several colleagues in my research community. Rianne Valkenburg provided me with her marked up copy of the Delft Workshop Protocols study data set so that I could compare my computational results with hers. Professor Petra Badke-Schaub and her research assistant Joachim Stempfle offered me their data set. Their intellectual generosity in sharing data has, I believe, helped to progress our field of design studies by allowing research instruments to be developed, calibrated, and experimented with on common data. I hope this will become the norm.

Support for my empirical research for this book has been provided by the Australian Research Council and the University of Sydney. I express a very deep gratitude to these institutions for, as I see it, taking a risk on a young academic.

This book integrates, reworks and updates several papers which I have previously published. Portions of the text in Chap. 3 appeared in the following papers: Dong A (2005) The latent semantic approach to studying design team communication. *Design Studies* 26:445–461. doi:10.1016/j.destud.2004.10.003; Dong A, Hill A, Agogino AM (2004) A document analysis technique for characterizing design team performance. *Journal of Mechanical Design* 126:378–385; and Hill A, Dong A, Agogino AM (2002) Towards computational tools for supporting the reflective team. In: Gero JS (ed) *Artificial Intelligence in Design '02*. Kluwer Academic Publishers, Dordrecht, 305–325. Portions of the text in Chap. 5 appeared in the following paper: Dong A (2006) Concept formation as knowledge accumulation: a computational linguistics study. *Artificial Intelligence for Engineering Design, Analysis and Manufacturing* 20:35–53. doi:10.1017/S0890060406060033. Portions of the text in Chaps. 2 and 6 appeared in the following paper: Dong A (2007) The enactment of design through language. *Design Studies* 28:5–21. doi:10.1016/j.destud.2006.07.001. Portions of the text in Chap. 5 appeared in the following papers: Dong A (2006) How am I doing? The language of appraisal in design. In: Gero JS (ed) *Design Computing and Cognition '06.* Kluwer, Dordrecht; and Wang X, Dong A (2008) A Case Study of Computing Appraisals in Design Text. In: Gero JS, Goel A (eds) *Design Computing and Cognition '08.* Springer, Dordrecht. I am thankful to the publishers for allowing me to incorporate these papers into this book.

Lastly, this book is dedicated to my family. I am indebted to the support of my mother Tu, my sister Anh, and our dog Norton. Thanks for making this struggle worthwhile. I am particularly thankful to my 'adopted' grandparents, my Nana and Da Mary-Alice and Bill Aultman. Nana and Da helped to sponsor our family from Vietnam and worked tirelessly to re-settle us in Pasadena, California. Nana even gave me the name of Andy when it was clear that my Vietnamese name was not going to be easily recognized as a boy's name in America. How do you express your gratitude for such generosity? I hope 'my' achievements can be seen as a reflection of theirs.

Contents

1 Designing and the Language of Design

> If we don't invent a language,
> if we don't find our body's language,
> it will have too few gestures to accompany our story.
>
> Luce Irigaray, *The Sex Which Is Not One*, 1985, p. 214

Rethinking the Designer

The Kurnell Peninsula in the state of New South Wales (NSW), Australia, lies about 35 km south of the center of Sydney. Its current reputation as the site of one of Australia's major petroleum refineries, sand mining, and the flight path for the north-south runways of Sydney Airport conceals its auspicious roots as the landing spot of James Cook in 1770. It is also the site of a design project that sets the stage for our approach to thinking about the operating principles that give rise to design as a work, as praxis, and as a field.

Australia's reputation as the driest inhabited continent, the driest continent is Antarctica, means that fresh water supply is a perennial concern for Australians. Mark Twain was referring to the Western United States when he wrote that "Whiskey is for drinkin' and Water is for fightin'", but his assertion could aptly characterize the debates surrounding sustainable water systems for Australia. Concomitant with a time when the dam level at the Warragamba Dam, which supplies 80 percent of metropolitan Sydney's drinking water, dipped below 40% and mandatory Level 3 water restrictions were applied, the New South Wales Labor Premier Bob Carr and the Department of Infrastructure, Planning and Natural Resources (DIPNR) released a metropolitan water plan titled *Meeting the challenges – Securing Sydney's water future* (NSW Department of Infrastructure 2004). In the report, the DIPNR set forth desalination as a contingency plan for dealing with Sydney's diminishing freshwater sources *among others* including recycling and reducing demand. In fact, desalination is presented only as "planning for desalination" (NSW Department of Infrastructure 2004, p. 12) rather than an emphasized alternative. Then, on 8 May 2005 (New South Wales. Parliament. Legislative Council. General Purpose Standing Committee No. 5 2006), Premier Bob Carr announced the selection of the Kurnell Peninsula as the site of a desalination plant. The announcement shocked and disappointed the community of Kurnell as no prior consultations had taken place with the local citizens regarding the location of

A. Dong, *The Language of Design*,
© Springer 2009

the plant (Sutherland Shire Council 2005), and the branding of the desalination plant as a 'major infrastructure project' meant that routine checks on the project could be bypassed per Part 3 of the *Infrastructure Implementation Corporation Bill 2005* (Parliament of New South Wales 2005)[1]. The beginnings of the account of the design of the Sydney desalination plant in the DIPNR report had the effect of bracing the authority to design in the government agencies rather than in the communities that the facilities service or in the designers of sustainable water systems.

It takes only the merest reading of the debates surrounding the design of a sustainable water resource for Sydney to know that the public was not, in the eyes of the NSW Government, authorized to design. In July 12, 2005, the *Sydney Morning Herald* quoted the NSW Planning Minister Craig Knowles as stating that the design of the Kurnell desalination plant is "beyond public debate" (Frew 2005). The debate was to be more than about the location of the desalination plant, but whether a desalination plan would be a desirable designed work in response to the brief – a sustainable water supply system for Sydney. There was even doubt as to whether that agency would exist in the designers of the Sydney desalination plant. Any agency in the actions of the designers of the Sydney desalination plant or capacity of the designers to design anything else but a desalination plant as the response to the design brief had been forcefully diminished in legitimacy. The account of the Sydney desalination plant articulates an un-negotiated version of designing that remains complicit with the tripartite of designers who design, the public (users) who provide input and consume the designed work, and the policy-makers (capital holders) who brace the authority to design to those schooled in design, the designers. The account implicitly privileges only some with the capacity to design[2].

---

[1] In 2005, the NSW Parliament passed the *Infrastructure Implementation Corporation Bill 2005* (IIC). In Part 3 (Functions of IIC in relation to major infrastructure projects) the bill authorizes the Premier to establish "project authorisation orders". "A project authorisation order may be made on such terms and conditions as the Premier determines and as are specified or referred to in the order." The *Environmental Planning and Assessment Act 1979* Sect. 75A specified the types of projects that would fall under the Premier's authority to establish "project authorisation orders". The act prescribes that "major infrastructure development includes development, whether or not carried out by a public authority, for the purposes of roads, railways, pipelines, electricity generation, electricity or gas transmission or distribution, sewerage treatment facilities, dams or water reticulation works, desalination plants, trading ports or other public utility undertakings." As such, authority to proceed with the desalination plant could vest in the Premier. In addition, the IIC authorizes the corporation to seize land "by agreement or compulsory process." Finally, as if the above were insufficient to vest complete authority over the design of major infrastructure projects, the act authorizes the corporation to bypass Sects. 34–37 of the *Public Works Act 1912* which sets forth routine checks and balances such as legislative review.

[2] As I wrote the first draft of this chapter, the NSW Government decided to jettison the desalination plant in February 2006 given public opposition to the plant. I optimistically thought that this would be a victory for citizens doing design. I was too optimistic. In January 2007, the NSW Government once again declared its intent to proceed with the desalination plant, and, rather rapidly, produced engineering reports and draft tenders. The instantaneous appearance of these documents would imply that design and engineering work on the desalination plant had continued

Moving across the Pacific Ocean, we find a contrasting picture. On October 17, 1989, at 5:04:15PM PDT, right before the first inning of the last game of the World Series "Battle of the Bay" between the Oakland-based A's and the San Francisco-based Giants, a magnitude 6.9 earthquake struck the San Francisco Bay Area. Centered near Loma Prieta peak in the Santa Cruz Mountains, situated south of the City and County of San Francisco, the Loma Prieta Earthquake destroyed significant infrastructure around the Bay Area, including collapsing the Cypress viaduct of Interstate Highway 880 near the City of Oakland, collapsing a west-bound section of the San Francisco-Oakland Bay Bridge onto the lower-level eastbound section, and severely damaging parts of the Central Freeway that directs freeway traffic from the East Bay onto the streets of San Francisco (and perhaps most memorably, at least for me, deleting my freshman chemistry report when power was cut). It is the redevelopment of the Central Freeway in San Francisco that we turn our attention to.

While the tale of the Central Freeway Replacement Project and related Octavia Boulevard Project is fairly typical of any development project in the politically active and aware San Francisco, complete with no fewer than three voter proposi-tions relating to the project[3], the story provides a contrast to the Sydney desalina-tion project. Originally, the California Department of Transportation (Caltrans) planned to re-build the damaged portions of the Central Freeway as it was. How-ever, it was widely acknowledged, given the social failures caused by the San Francisco Redevelopment Agency Western Addition Project A–1 which widened Geary Boulevard and arguably decimated the African American community in the spirit of 'urban renewal', that the re-construction of the Central Freeway offramp would continue a legacy of environmental racism toward those least able to chal-lenge projects. These past projects were largely determined beyond their control. While a complete account of the path by which the Octavia Boulevard Project proceeded would constitute a book in itself, the result of the project is widely acclaimed as a success in terms of a project which could forthrightly be character-ized as being designed-at-large by the community. In direct contrast with the

---

despite public proclamations otherwise. The Sydney desalination project is proceeding as of 2008 with contracts awarded to two private joint ventures, Blue Water Joint Venture, which comprises of Veolia Water and John Holland, and the Water Delivery Alliance, which comprises of Bovis Lend Lease, McConnell Dowell, Kellogg Brown & Root, Patterson Britton & Partners and Environmental Resources Management.

[3] First, voters adopted Proposition H in November 1997 which authorized Caltrans to rebuild the Central Freeway with a four-lane, single-deck, elevated structure over Market Street from Mis-sion Street to Fell Street. Then, in the November 3, 1998 General, voters repealed Measure H with Measure E, The Central Freeway Replacement Project of 1998, in which they voted in favor of the question, "Shall the City repeal 1997's Proposition H and authorize Caltrans to replace the Central Freeway with an elevated structure to Market Street and a ground-level boulevard from Market along Octavia Street?" Finally, in the November 2, 1999 Election, Proposition J Central Freeway Replacement, voters approved the measure asking "Shall the City repeal 1998's Propo-sition E and authorize Caltrans to retrofit and widen the lower deck of the Central Freeway to provide a four-lane, single-deck structure over Market Street from South Van Ness Avenue to Oak and Fell Streets, and prepare an annual transit plan?"

Sydney desalination plant, the design brief itself was in many parts written by the community, solutions were proposed by various members of the community, those solutions were debated and reviewed (and voted upon by all citizens of San Francisco) without legal hindrance that would fully brace the authority to design in others.

In accounts of design, we are lead to a quandary, then. When a designer or designers give an account of designing, who is the 'I/we' who is doing the designing? Books and popular magazines that pen veritable homage to designers, while usefully cataloging inspiring and innovative designed works, depict designing according to presumed truths about what constitutes design. In the predominant models of designing, such as the cognitivist (Lawson 1997) and information-processing (Rowe 1991) paradigms, it is thought that what is displayed as designing is the product of the designer's cognitive activities. Process-oriented models of designing such as user-centered design reinforce the notion that designing is to be understood in terms of a center-periphery model in which the designer receives inputs and gives an output, the designed work. This model of designing offers an account of designing that retains the stability of an identity of a designer as a kind of person; those who provide input but who are not directly involved in conception are not the designers. Normally, the others are referred to as the stakeholders. In other formulations, the designer is the conceiver; the user is the receiver. These conceptualizations of designing also maintain the stability of the concept of consumers of designs and producers of designs. Any other account of designing does not appear to be like designing; a designer is only a designer to the extent that the person 'realizes' the legitimate enactment of design.

I would like to divert here slightly to a literary work which delves into this issue on the construction of the identity of a 'creative worker', in this case a writer instead of a designer. I think this diversion may clarify this line of questioning about the identity of designers and the (agency of) actions by designers. In the book *The Master*, Colm Tóibín tells the story of the introverted life of the author Henry James and the reference points, such as 19th century London, around which Henry James organizes and adopts his writer's identity and his concomitant writing behavior. Tóibín tells how James, in committing himself to a reclusive life of writing, silently submits to the forces that shape James's life and his writing. Tóibín relates these forces in the following passage in which James begins to write a fictionalized account of his encounter with a Russian princess in Paris:

> He took the pen and began. He could have written an indecipherable script, or used a shorthand that only he himself would understand. But he wrote clearly, whispering the words. He did not know why this had to be written, why the stirring of the memory was not enough. But the princess's visit and her talk about banishment and memory, of things that were over and would not come back, and – he stopped writing now and sighed – her saying the name, saying it as though it were still vividly present somewhere within reach, all these things guided his tone as he wrote. (Tóibín 2004, p. 9)

While it is James who writes (i.e., performs the action of writing) the princess's words and James who gives her words the air of reserved exasperation for which he is known, Tóibín invites us to consider how the forces that shape his identity also shape what he does and how he does his writing. This is a question that the philosopher Elizabeth Grosz has taken up in conceptualizing the forces that shape action:

> The interesting question is not who am I, what am I, how am I produced, or how is my identity stabilized – although these aren't irrelevant questions. The more interesting question is how do I act, what enables me to do this, what acts in me when I act? And in switching to the question of acting from the question of identity is a powerful shift. It's a different way of understanding how we organize, what in us is organized, whether we require a plan, and whether we require a certain intentionality. (Ausch et al. 2005)

For James, it was never apparent whether or not he had a plan, an "intentionality" for his life. Tóibín imagined James being asked about his life by a sculptor Hendrik Andersen who maintained a restrained intimate relationship with James:

> But did you not once plan it all? Did you not say this is what I will do with my life? (Tóibín 2004, p. 310)

Thus, the question that is raised in relation to design is the extent to which we can ascribe design to the perspectival account (testimony) of the designer. The model of the lone designer designing is being challenged by citizen groups and ordinary people engaged in online and often collaborative digital media design and production. Increasingly, design is held in the multitudes rather than in a designer or designers, that is, people to whom institutions have conferred the authority to design and to be called designers. And when designing is held in the multitudes, what 'acts in a designer', to borrow the words of Elizabeth Grosz, is not necessarily proximally linked to any one designer or even several designers. Let us examine several examples in which the identity, actions and behavior of the designer are enveloped by these 'influences'.

User-centric design processes such as user-centered design (Vredenburg et al. 2002), empathic design (Leonard and Rayport 1997) and contextual design (Beyer and Holtzblatt 1999) hold as their primary philosophy that the consumer of the designed work is the spark for new product ideas, new designs. Gilmore and Pine (1997) identify collaborative and adaptive approaches to mass-customization in which the consumer engages with the designer in customizing a product. In these approaches to designing, there is a forthright acknowledgement that the designed work emerges from direct interactions among the designers and the consumers of the designed work. The designer channels the user's insensible and invisible desires toward the conception of the designed. What makes the designed work sensible and visible is the innervation of the designer to the conditions for designing grounded in the user. The user is, in user-centered design processes, partially implicated in the basis for actions by the designer. While the user-centered design model still maintains the user at the periphery of design, this model points toward the fallacy of not regarding the user as a constitutive element shaping the identity of the designer and the actions of the designer.

In Mumbai, successful efforts in community involvement turned slum dwellers into property developers (Mukhija 2003). An imperative for the government agencies, private equity holders, and slum dwellers to act collectively resulted in several successful redevelopment projects. As such, the slum dwellers were the designers, consumers, and financiers. Such projects are not isolated and one-off; in fact, they are increasingly becoming the norm in poverty reduction development projects. The UN Millennium Development Goals makes an explicit policy statement about expanding the capabilities of the urban poor to design their environment. Reflecting on a project for capacity building of women in Mumbai to design better settlement, the Society for the Promotion of Area Resource Centres (SPARC), an Indian NGO, wrote, "SPARC and its partner organizations have demonstrated that pavement dwellers have the capacity not only to save to provide for their housing, but also to provide tangible inputs into the design and construction of their homes." (Society for Promotion of Area Resource Centres 2005) One more example that deserves mention at the same time as discussing these collaborative, community-driven design practices is the explosion of personal digital content creation. In 2006, *Time Magazine* anointed "You" as the "Person of the Year 2006". Citing YouTube, MySpace, and Wikipedia as examples of people designing, producing, and publishing their own content, often collaboratively as in the case of Wikipedia, the Editors of *Time* proclaimed that the phenomenon of personal content creation is "a story about community and collaboration on a scale never seen before." I would add design to their list.

In industry, the design of large-scale systems requires the efforts of very large teams of designers. Take for example the design of the much-heralded Airbus A380, which is designed by a consortium of 4 European aerospace companies forming the European Aeronautic Defense and Space Company (EADS) and BAE Systems. The core team, if you will, does not take into account a worldwide contingency of suppliers and multinational airlines as customer-designers. Boeing's 787 Dreamliner project is unprecedented in the amount of design and production sub-contracted to other companies. Using an integrated CAD software package with all of its suppliers, Boeing was able to sub-contract significant portions of the design and manufacturing of the Dreamliner to suppliers such as Alenia Aeronautica, Mitsubishi Heavy Industries Ltd, Kawasaki Heavy Industries Ltd, and Fuji Heavy Industries Ltd. The traveling public and curious could login to the Boeing 787 marketing Web site www.newairplane.com to join in the design process as part of the World Design Team.

The actions and behaviors of the individual designers need to be accounted for within a set of pre-conditions that inscribe a limit on the actions of any one designer as a transformer of forms; as such, frameworks which, in the words of Rowe "fashion a generalized portrait of design thinking" (Rowe 1991, p. 1) may need to come to terms that what is observed as designing may actually just be an enactment of choices already made preceding the designer. Designerly actions and behavior may instead have arisen from operating principles that serve both to constitute and relay design praxis. All of the examples of design above might share a blend of these operating principles but to different degrees and with different foci. From these operating principles, designers construct a revisable and reconfigurable design praxis which characterizes their design identity and contributes to the identity of their discipline.

Norms regulating design praxis are often formally and overtly expressed; they exert a force on how designers describe design. Quotes from prominent designers (architects) such as "Less is more" (Ludwig Mies van der Rohe) and "Form follows function" (Louis Kahn) circulate design cultures. They situate accounts of design to the body of design knowledge and normative practices which regulate the identity of the designer as well as the acceptability of the designer's work. The designer is effectively reiteratively performing a design practice. By affixing accounts of designing to a particular way of designing, design practice can be named according to the historicity of norms as both coherent and ontologically describable.

In each of these examples, the agency to design and the available actions to the designer operated as an intermediary between the designer and the designer's matter and methods. Who the designer is and what the designer does are relevant, but the operating principles that produce the designer's identity and regulate the designer's behavior erected forms of mediation through which design is enacted. Clearly, the designer or group of designers is implicated in designing, at minimum in forming a mental representation of a materializable work which others follow through in order to make the design. However, the boundary between the designer and the others and the boundary between being in/within the process of design and outside of what is considered designing is artificial. Delineating boundaries between what is designing and what is not designing reinforces and undergirds the division of labor that gave rise to design as a commercial practice and remunerable profession. The modern designer stands in contrast to the artisan-crafter who both 'designed' and 'made' a work. While this delineation suits an efficient and rational system of production through the creation of capital value, maintaining and institutionalizing this division makes it 'natural' that design is ascribable to a person and that design is prescribable, that is, that there are normative ways of 'doing' design. In the example of "You" as the "Person of the Year" the implication is even more far-reaching; anyone could design. So, does the concept of a 'designer' even exist anymore?

Let us pull back for a moment and consider some objections to this line of thought. Opponents to these thoughts should at this point respond that collaboration is already an active component of any design project, and that design is inherently collaborative. That design is collaborative is an essentialist's position. Certainly, there are barriers to collaboration in design, which have been widely studied, but the problem is not that designers never collaborate. Collaboration in participatory design makes users *res ipso* designers (Carroll 2006). I concede to this argument, but I would question the codes of conduct which regulate the collaboration. In many cases, collaborative design is more about amelioration than it is about genuine joint activity. Likewise, though with fewer social justice ramifications, saying that an architect collaborated with a structural engineer on a highrise skyscraper project is prudent, but the alternative is to claim design monasticism, which is unrealistic.

Who the 'we' are who participate in design matters not simply as a concern for design theory or for collaborative design or for academic philosophy. It is not sufficient to claim that if we get the mechanics of collaboration such as 'town halls' (Innes and Booher 2004) and online discussion forums right, then we have licked the problem of collaboration in design and might even go so far as to claim a cooperative model of design. If those who are doing the designing are also those who are claiming the model of designing and claiming what counts as collaboration in design, then the inmates really are running the asylum, to borrow from the title of the book by Alan Cooper. Thus, I am skeptical, though not cynical, of claims to the 'we' bandied in discussions of design collaboration. While I have no doubt as to the truthfulness of the claim to collaboration in design, the issue is about the degree of remarkableness, importance, and interest in the 'we' who speak about collaboration. One could simply say that poor design collaboration has to do with nothing more than exclusion, power play, and politics. I would argue otherwise; the problem might lie in the operating principles for design and which principles are emphasized in the enactment of design.

One other criticism to my argument would be, are people who are contributing to YouTube, MySpace or Wikipedia really designing? Does their existence really spell the death of the designer? The knee-jerk 'No' response might be that they are not designers so none of this really matters; in many instances, they appear to be modifying pre-existing templates previously designed by a graphic designer. However, many designers routinely modify designs; parametric design and routine design are words used to describe this kind of designing. So, faulting these personal content creators as non-designers by virtue of not innovating something entirely 'new' is rather facetious. However, an anything goes definition of design is not palatable either. My response is that they are designers; they are just operating with a different set of codes which emphasize the operating principles of design in a way that differs from the 'rules of the game' practiced by professional (i.e., paid) graphic and multimedia designers.

And what are these operating principles and what do they have to do with design as such? Am I just trying to catalog new essentialist attributes of design? Again, I would argue otherwise. The examples that I have pointed to above, I believe, are tied to the specific ways in which design is enacted by various disciplines and by the way that design as a field organizes itself. I believe that they are tied to the various preconceptions that people have about the social codes of design praxis. Above all, the examples of the clashes that occur in design collaboration, the 'new' models of community-based design, and 'anyone can design' euphoria of Web 2.0 applications fit with the major empirical fact of the marked growth of design as profession and populist design-like behavior.

Design, as a work, as praxis, and as a field has been radically altered by the examples that I have drawn above. The forces of commoditization of design react with the forces of the exultation of design and the deification of designers. The forces of the democratization of design push up against the established powers that recognize what is/are legitimate design practices. Theories of design that rely on some sort of scientific or innate human behavior in design, which link design, science, rationalism and technical methodology, inevitably miss the circulation of mass images and messages about design that are (already?) discursively producing new enactments of design.

At the same time, I am not so pessimistic as to believe that the only concept of design that is possible is that of the camel, infamous as the image of the horse designed by committee. Instead, I would propose that design is an enactment of a set of operating principles wherein the actors emphasize different aspects of these principles. It is the varying emphases, which probably exist for valid reasons, on these principles that lead to various design praxis. What is desired is a framework for thinking about design that is capacious enough to deal with the ways in which design is enacted. We would like to discover a set of operating principles of design that act as a set of forces providing resources for defining design identities, design praxis and the design field.

In order to deal with as diverse a field as design, with all of its visual languages and dialects, it is the discourse of design, the language of design, which we turn to for these principles. Turning to the language of design raises the fundamental question: how can language design? To understand the perspective that I will take to address this issue, it is first important to acknowledge the historical ways in which understanding design has been dealt with and which inform my perspective.

As I stated above, the interest in this book is to question what constitutes the productive forces of the "performance of innumerable semiotic acts" that go into the accounting of design and the designed work (Medway and Clark 2003, p. 270) and how those operating principles are capable of producing an effect, the designed work. This issue has formed the basis of an important area in design research known as cognitive design research. Cognitive studies of design are part of a larger endeavor to study design scientifically. Herbert Simon (1996) famously conceived of design as a "science of the artificial", the idea being that design could be studied scientifically and, more or less, practiced scientifically.

Put simply, cognitive studies of design investigate ways of thinking in design. The cognitive structures of the individual or group mind while designing are the basis of design theory. The problem with this perspective is *who is looked to* and *what is looked at* as evidence of design thinking. Is designing 'natural' in the sense that humans are genetically programmed to design and hence all we need is to uncover what is programmed into the mind that enables humans to design? If so, it matters not who we look to; everyone is designing as the practice of everyday life. Or, is designing "artificial" according to Simon: "The intellectual activity that produces material artifacts is no different fundamentally from the one that prescribes remedies for a sick patient or the one that devises a new sales plan for a company or a social welfare policy for a state." (1996, p. 130). In making this case, Simon is actually stating that despite design's 'artificiality', it can nonetheless be studied scientifically by looking to design professionals. Finally, might we argue that designing is 'performative' – producing what it describes through intentional behaviors that cite prior practices to make the act ontologically describable? If this is the case, then we must look back at ourselves, the observer, and interrogate what we expect to observe in the 'designer', which alternatively questions the autonomy of the designer detached from ontological expectations of designing.

I take this third position. The organization, construction and transformation of mental representations should be seen as party to the norms which regulate the conduct of design practice. The rationale, intent, influences, and ideas provided in a designer's account of design should be seen as linguistically enacting the designed work, not just describing it.

This orientation toward design should not preclude cognitive models of design nor does it sublimate 'schools of design' and their influence on designers, design practice, and designed works. However, I believe that there is more to the picture than lumping these as influences on the designer. The act of designing is not just a semiotic act of imbuing a form with function and meaning, but, rather, the effect of assemblages of prior practices that act through the designer to make a designed work. The assemblages of practices are a condition of possibility for a designer's cognitive behavior. 'Schools of design' provide a face for one of the vectors of practice and set up a particular language by which to express designed forms. Users and users' input provide a face for another vector of practice. And so on. The point is that while the way that a designer thinks, the way that a designer operates tools and media, and the observed modes of practices by which a designer works are indissociable aspects of designing, they must first be understood as 'things that happened' on the way to becoming design.

It is this journey to becoming design and the emergence of the designed work that provokes us to turn our attention toward the driving forces that shape a designer's identity, the designer's behaviors, and the designed work. Donald Schön set forth a classic metaphor of designing as a reflective "conversation with the materials of the situation" (1983, p. 78). We could take this metaphor of designing as a conversation one step further to designing as a performance. If we take the concept of performance in its most general sense, namely the production of a subject through the performance, then design practice and the designed work are the effects of a performance. The conversation with ideas is part of a series of dialogues in this performance. But because language is a part of what designers say and write about what they do, what is the role of language in the performance of design? Does it merely give expression to and foreground design activities? My contention will be that by thinking of language use in design and its semantics and grammatical structure as 'doing' design, the linguistic descriptions become more than representational bystanders that accompany actions. They become reality-producing performatives. If we could understand how design has been 'done' by language, we can then begin to question the extent to which design exists within language. We thus turn our focus towards accounts of designing through the language of design.

Design and Language

Design is sometimes described as a 'science' of the world the way it could be instead of the world as it is. It may seem odd to suggest that language vis-à-vis words 'designs'. Yet, the act of writing down ideas into a design journal or into an online blog accomplishes several goals. Semantically labeling a design concept with a word assists with the recall of the concept from memory at a later time. Creating the semantic label for the design concept also allows designers to 'think by writing' in ways similar to 'thinking by sketching'. Through interactions with design representations, the designed work is at once realized but also made mental in the designer's mind.

It is to these texts that designers produce during designing that people turn to for insight into the design process and the designed work – to get at what the designer was thinking. Richard Buchanan theorizes that design actually involves a "skilful practice of rhetoric ... through all of the activities of verbal invention and persuasion that go on between designers, managers and so forth, but also in persuasively presenting and declaring that thought in products." (1989, p. 109) Accounts of design appear in design texts. For brevity, we refer to what can be conversed and written about design as design text. Design texts come in a variety of forms, from informal correspondence to communiqués. The language of design texts serves constitutive, constructive and instrumental roles toward design. As its constitutive role is the focus of this book, let us first turn to its representational, instrumental and constructive roles.

Let us start by examining how language serves as a vehicle to represent the designed work and to describe the design process. One convenient way to think about language's representational role is to think of relating design as a story telling process. In the story, the designer is the protagonist (or the designers are the characters), the process is the plot, the tools are the props, and the product is the theme. What was designed and how the design process proceeded is an 'end' of the story. Story telling in design is about establishing a coherent narrative about the design process and the designed work, in a sense, rationalizing the design process. Peter Lloyd writes that story telling plays a role as a mechanism for relaying the social experiences and for the "development of a common language" (Lloyd 2000, p. 357) in design. Story telling fixes the dialogue of pluralistic contributions by those involved in the design process into the single designed work, to bring "harmony to the underlying object world" as Bucciarelli (1994, p. 187) wrote. The stories generated explore various facets of the design process and the designed work. They may convey the value the product will bring to the people who will use it. They may capture the real-world context in which the design concepts evolved during the design process. The 'plot' could reflect the conflicting interests and resulting reconciliation and shared agreements of the design stakeholders. Design practice could make for good television. Research in design as story telling shows that there are as many ups and downs in professional design practice as your standard television drama. Viewing design as story telling frames design practice in an emergent perspective to understand how collective properties of design – technical problem solving, social networks, information processing – contribute to a narrative of the design process and designed work. There are story telling archival systems employing multimedia and moviemaking metaphors to capture the description of the design and enrich the textual explanation of the designed work (Garcia et al. 2002).

The stories about design that people know, which design stories to relate, how to tell them, and how designers describe their practice constitute a common sense that anchors designers to a body of knowledge and practice. These stories have a commercial market. People turn to texts about design, such as the magazine *Wallpaper*, or a book on Michaelangelo's architectural style, such as *The Architecture of Michelangelo* by J.S. Ackerman, to appreciate designed works and to experience designing. Reading a legal brief, novel, or news article provides none of the pleasures associated with flipping through the Tauschen series of books about designed works. Clearly, people want to know more about a design than just to appreciate the work through the senses.

To a first-approximation, then, we could say that the linguistic content (words) of design texts provides a useful index to the composition and structure of key design concepts. Based upon this proposition, we might propose the following baseline theory of the language of design:

> A theory on the language of design – The linguistic content of design text is related to a conceptual structure of the designed artifact, and is produced and represented through the lexicalization of concepts.

That is, the purpose of language use in design is to establish a set of ideas about the designed work; the text, whether authored individually or collectively, is a written expression of design advocacy. The premise is that the design specifications and solutions as communicated through the design text are functionally related to a conceptual model of the designed work. Certain combinations of the chosen properties of the designed work gave rise to the corresponding combinations of design descriptions in the design text. The connections between the different parts of these descriptions in language explicitly express the designers' conceptualization of the designed work.

Such a representational theory of language use in design is useful for studying how designers think. An empirical study of design documentation and design conversations could elucidate how the representation of designs through words provides a viewport into the way that designers think about their artifacts. Aspects of design thinking such as knowledge sharing and bridging between designers and the influence of emotion on design thinking could then be revealed through a close examination of the language. Studies based on a representational theory of language use in design use language-based communication to develop a model of how designers think (Kan et al. 2007; Stempfle and Badke-Schaub 2002).

Its instrumental role in design, particularly as a tool of communication and of recording design intent and rationale, is language's most functional role. There really are no surprises here. One of the more ambitious projects to turn language into an instrument of design occurred in the early 1990's. The Center for Design Research at Stanford University and the company Enterprise Integration Technologies (EIT) developed a language called KIF (Knowledge Interchange Format) so that designers could communicate and transfer knowledge among different "knowledge bases" residing at various companies, requesting information and services from each other (Cutkosky et al. 1993). While the system met reasonable success, the main finding was that the language facilitated design coordination – that is, the function of the language was not just merely for the representation of an artifact but also a mechanism for distributing working and design information sharing. Today, information technologies such as e-mail, teleconferencing, and product data management systems create media and channels for communication among designers. These communication media and channels also define a social network developed among the communicators, transmitting their ideas and their shared interests.

When we mean that language is constructive in design, we mean that the use of language during design is effective in contributing to designing itself rather than as a support function. Again, there have been numerous intriguing studies demonstrating language's constructive role. The technique of 'brainstorming' is a commonly-practiced cold start idea generation process in which groups of people propose (potentially outrageous and infeasible) ideas by verbalizing them. The idea behind brainstorming is that the energy generated by the verbalization of the widest possible space of design ideas by a group of people may lead to the creation of novel solutions, though empirical evidence as to the actual effectiveness of brainstorming is inconclusive. Mabogunje and Leifer (1997) offer empirical evidence to suggest that the creation of more unique noun phrases that label design concepts during the conceptual phase of design correlates with quality design outcomes. Bauke de Vries and colleagues at Eindhoven University of Technology (2005) developed a system called Word Graphs to stimulate the thinking of architects. Architects were placed in a semi-immersive working environment in which their horizontal working surface consisted of a digital drawing space (actually a digital tablet) and information projected onto the periphery of the workspace. One piece of information is the Word Graph. As the architect is working, the architect is asked to speak out loud or write down words representing the ideas being generated. The Word Graphs system tries to display new intermediate words that semantically connect two previous words expressed by the architect. Suppose the architect was to make a statement such as, "The shape of the pavilion's roof resembles a leaf." Then, the Word Graph system displays a key phrase in the statement "leaf". Later, the architect might make another statement such as "I would like the building's structure to be organic, to remind the occupants of a plant". Then, the Word Graph system might display a semantically related term such as "flower" or "root". The prototype Word Graphs system operated as a sort of 'Wizard of Oz' system due to the limitations of speech recognition systems. By producing associations between words as labels for design concepts, new ideas were spurred and periods of inactivity (possibly designer's block) were reduced.

The stimulation of creative design thinking using words begs the question how it is that words can become part of the information processing operations involved in forming new design concepts. How is it that words *alone* assist designers in concept generation? In an intriguing study (Chiu and Shu 2008), researchers at the University of Toronto set out to investigate how designers use words as stimuli to assist in concept generation. The participants were given design problems. Along with the design problems, they were provided with stimuli in the form of words (verbs) related to the design problems. The engineering designers selected a set of those stimulus words as the basis for the formation of their design concept. For example, one of the design problems required the engineers to design a method to remove the shell from sunflower seeds so that the oil could be extracted. Stimulus words included rinse, evacuate, eliminate, cut, and remove (the base word from which stimulus words were located). The researchers found that words that were slightly more abstract than the base word were the most useful to the engineers in formulating a complete design concept. Thus, words such as cut, wash, and flush were useful in stimulating a process of conceptual combination, the integration of two (or more) concepts, into the design concept. The most innovative solution involved the most abstract term, flush. In this solution, the engineer proposed a concept which involved first fracturing the shell, then pressing the entire shell and seed for oil without extracting the seed. The oil could then be 'flushed' out of the fractured shell.

It is possible to conjecture that the stimulus words caused, in terms of cognitive processing, the production of a thematic relation between the stimulus (verb) and a possible thing (the designed artifact) using only the principle notion of the word with a more or less arbitrary association to the artifact. This type of association is common in English. If I gave you a combination of two nouns such as *pill box* or (slightly more irregular) *time flies* you would likely state that the *pill box* is a *box* used to contain *pills* and that *time flies* are *flies* that travel through *time*. The relation between these words is omitted, but it is something that you can supply. It is possible that the same phenomenon is going on in the previously cited study. The designer creates the relation between flush and the designed artifact. Here, it is a noun–noun combination on the order of *seed flush*, a *flush* of oil from a *seed*, and it is the artifact that realizes the device for *seed flush*.

The noun–noun combination *seed flush* is a type of combination which displays the creation of a new representation that is linked to the original modifier (*seed*) and head noun concept (*flush*). In such a type of combination, we are not filling in a slot for the schema. With the combination *pill box*, it is known that a box (the schema) is used to hold one or more objects (slot). It is to be expected that if the modifier is *pill*, then the word *pill* fits into the slot for schema of box. However, noun–noun combinations such as *time flies* and *seed flush* are more complicated. They require us to infer, or in the case of design to envision, additional character-istics of the head noun concept that are not directly indicated by the modifier. Christina Gagné (2000) conducted studies on the ability for humans to draw con-nections between noun–noun combinations. Gagné found that humans find it eas-ier to draw relation interpretations than property interpretations. That is, it is easier to envision a plausible relation between *seed* and *flush* than it is to create a prop-erty similarity between them. Franks and Rigby (2005) theorized that the ability to draw property relations between unusual noun–noun combinations, such as those of the variety like *time flies*, is likely to be associated with a display of creative thought and is a skill that might have been evolutionarily conserved. The display of creative property-mapping interpretations of noun–noun pairs, they believe, is likely to be linked to sexual display for evolution purposes of picking a smarter mate. Picking a smarter mate increases the likelihood that 50% of your genes will be passed on to an offspring with a strong chance of perpetuating your genetic material to your grandchildren and so on. They used this assertion to explain their findings on property-mapping interpretations of noun–noun relations. They found a sex-based difference in the creativity of language use. The participants in their study were presented with unusual noun–noun combinations and asked to produce a definition for them. The definitions were then categorized as either being a property mapping or a relation mapping. In the twist to this study, the experimenters were either attractive young females or attractive young males and participants were motivated to display creative language use in the presence of experimenters of different sexes. They compared the difference in performance to the baseline condition and found that males produced significantly more property relations than females, which they believe arose as a consequence of the males' motivation to display linguistic creativity in the face of the attractive, female experimenters. So much for knowledge as motivation for research!

The representational, instrumental and constructive views are nonetheless partial. Seeing language as purely representational, for example, misses the point that accounts of design through language implicate language with 'doing' design. Language is more than representational, more than standing in for a design concept when no other representation yet exists, more than being a pointer to places in the mind to assist in constructing memory about a design concept, more than a passive historical account. Otherwise, language is impossibly estranged from what it is deployed to do – enact design and realize the designed work.

Certainly, I do not intend to reduce design to a set of linguistically formulatable propositions nor wish to imply that every aspect of design or a designed work could be characterized by words. Certainly, designing is a language on its own, partly performing what cannot be conversed or said but only enacted through designing and the designed work. Nonetheless, language gives expression to the becoming of design. Design text is not imitating designing, but rather amassing together the constitutive elements of design. While stating, claiming, and naming what is design and what is the designed work are not enough to effectuate them, these linguistic proclamations are important in the actualization of design.

At this point, some readers may be thinking that our line of argument about design text giving life to an abstract concept, design, and the materialization of the designed work are akin to a Deleuzian pure event. Indeed, there does appear a parallel in what language does to designing and language's effect on events. Deleuze claims that pure events are singular, incorporeal entities, expressed in particular configurations and movements of bodies, always remaining indeterminate and open-ended – until language is used. Language use involves the attribution or effectuation of the incorporeal events; the illocutionary designates the state between the incorporeal and the corporeal transformation. Likewise, the language of design gives expression to the becoming-design. Design, which was a set of acts which were indeterminate as to whether they would eventuate to a designed work and which were not yet ordered and ontologically describable as design, become enacted by language.

These cited design studies are thus acknowledgments that language is a part of the 'doing' (designing) and the thing 'done' (the designed work). In light of these accounts, we need a broader and more capacious account of the interconnections and complementarity of the constitutive, representational, constructive and instrumental roles of language in design. How do the activities, methods and practices of designing operate to constitute the materiality of the designed work through language? What are the linguistic resources that produce design practice and the designed work? The point here is that the semantic and grammatical forms of language use in design are not consequence free; they reflect the meaning potential of the designed work that is intended by the designer and the stakeholders. As an active functional instrument of design, language also becomes a constituent of design. By being active in reality-producing activities such as concept generation, language implicates itself as part of enacting design.

What we need then is a way of thinking about the language of design that propels us toward the main thesis of this book – that language is constitutively involved in the enactment of design. Simply put, we need to think of the language of design as *performative*.

The term performative traces its origins to speech-act theory. J.L. Austin (1962) named words with reality-creating potential as *performatives* when he delivered the William James Lecture at Harvard University in 1955. Austin distinguished between constative locutionary speech acts which utter, illocutionary acts which accompany actions, perlocutionary acts that implore actions, and performative speech acts which effect actions. Unfortunately, what troubled Austin was that in trying to clarify what constitutes and does not constitute a performative utterance, he eventually notes that every speech act is both constative and performative. In a sense, he recants his ontological ground for performativity.

Eve Kosofsky Sedgwick introduces the term "periperformative" (2003, p. 5). Sedgwick describes periperformatives in the following way. "Periperformative utterances aren't just about performative utterances in a referential sense: they cluster around them, they are near them or next to them or crowding against them; they are in the neighborhood of the performative." (2003, p. 68) Her discussion of the periperformative through their appearance in literature works to unmoor the Austinian notion of performativity – utterances that produce through naming, distinguishing between explicit performatives and non-explicit performatives – toward the ground of what she considers Austinian performativity actually concerns with. "Austinian performativity is about how language constructs or affects reality rather than merely describing it." (2003, p. 5) That is, the concept of language as creating reality – where the reality dealt with in this book is design practice and the designed work – is the axis of inquiry for performativity.

This ground has been most forcefully propelled by Judith Butler in her theories on gender and performativity. I will defer discussions of Butlerian performativity and its relation to the thesis of language enacting design to relevant portions of Chap. 2 and Chap. 4. For now, let me use the following account to sketch out briefly what will be termed the *performative operators*. The performative operators are the reality-producing linguistic patterns in the language of design. They are the fundamental components of the language of design that realize design in describing design.

J.S. Ackerman's quote of Michelangelo's 'theory' of architecture contains linguistic examples illustrating the performative operators acting together:

> When a plan has diverse parts, all those (parts) that are of one kind of quality and quantity must be adorned in the same way, and in the same style, and likewise the portions that correspond [e.g., portions in which a feature of the plan is mirrored, as in the four equal arms of St. Peter's]. But where the plan is entirely changed in the form, it is not only permissible but necessary in consequence entirely to change the adornments and likewise their corresponding portions; Whoever has not been or is not a good master of the figure and most of all, of anatomy, cannot understand anything of it. (Ackerman 1986, p. 37)

First, Michelangelo sets up a frame for the situation to which a specific design practice holds through an interrogative or questioning form of language: "*When* a plan" and "But *where* the plan". We will call this linguistic pattern of creating a frame *aggregation*. Next, he references this frame to create a materiality for the concept. The material of the concept is expressed through linguistic technicalization. He employs a field-specific meaning of the term "arms", as in the "four equal arms of St. Peter's", almost in a metaphorical sense rather than in a vernacular meaning. This is a process we shall name as *accumulation*. Finally, he valuates the concept by sprinkling expressions such as "not only permissible but necessary" and "a good master of the figure", a pattern we will call *appraisal*. Aggregation, accumulation and appraisal are the three linguistic patterns constitutive in producing design as the language of design describes design.

The way that Gaston Bachelard defines space in *The Poetics of Space* offers a more conceptual view of these performative operators. Aggregation is about producing a conceptual space for the designed work. Bachelard describes this as:

> Now my aim is clear: I must show that the house is one of the greatest powers of integration for the thoughts, memories and dreams of mankind. The binding principle in this integration is the daydream. Past, present and future give the house different dynamics, which often interfere, at times opposing, at others, stimulating one another. In the life of a man, the house thrusts aside contingencies, its councils of continuity are unceasing. (Bachelard 1994, p. 6)

The "house" is the frame of the design concept. It is the accumulation of past and present within the surrounding which gives potential meaning to the future (the designed work to come). The materiality of the accumulation is the place:

> in which we have experienced daydreaming reconstitute themselves in a new daydream, and it is because our memories of former dwelling-places are relived as daydreams that these dwelling places of the past remain in us for all time. (Bachelard 1994, p. 6)

It is through these "daydreams" that we appraise desires for another "world":

> Daydream undoubtedly feeds on all kinds of sights, but through a sort of natural inclination, it contemplates grandeur. And this contemplation produces an attitude that is so special, an inner state that is so unlike any other, that the daydream transports the dreamer outside the immediate world to a world that bears the mark of infinity. (Bachelard 1994, p. 183)

The metaphysics of space that Bachelard imagines is not only a way to conceive space but to map the "very extravagance of the being given to words" (Bachelard 1994, p. 114). This is our project.

The question that we must deal with is the connection between the ways that design is accounted for in language, such as in these design texts, such that language comes to inhabit design, such that language becomes design. That is, we are looking for correspondences between language use, including the way (grammar) that design text is expressed and the semantics of the design text, and the productive forces of language.

In order to validate our thesis, we would need to demonstrate, through analysis of design text, that the text enacted design, resulting in a designed work. Ideally, we would have a 'magic box' which could understand any type of language use such that we could easily convert words into data such as is done with verbal protocol analysis. Then, this analysis would be done – there would be some empirical data, some numbers and pie charts that we could show to illustrate our point. However, to attempt such an endeavor, given the human effort required, makes the method impracticable. While it may be possible to scale-up methods that make use of language to study human behavior in design from cognitive science and to study the productive forces of language from linguistics, it is not obvious how this could be done. The main impediment to explicating our thesis is the lack of suitable tools. It is for this reason that we turn to computational linguistics as our methodological approach. We cannot rely on computational linguistics to do verbal protocol or discourses analysis for us. That full natural language understanding is beyond the reach of computational linguistics also makes that effort futile. It may not even be desirable.

As is often the case when forging into new territories, we must try another approach. Our approach is to combine the theoretical and methodological rigor of 'science' with its emphasis on empirical data and repeatability of experiments and the intellectual and conceptual possibilities of critical theory. Science and critical theory work in tandem: critical theory gives us the viewpoints for thinking about design while computation allows us to interpret and materialize these viewpoints on data sets that present characteristics of design. My point is that science is not the only mode to study design nor should design be made into a scientific discipline for methodological expediency. Nor should we strive to formulate an approach to understanding design that privileges one over the other, science over critical theory, or critical theory over science. Rather, our mode of inquiry should broaden the informational basis of critical theory while enriching (computational) science with the social values and abstract possibilities of critical theory.

This approach will become more evident as we progress. The essential aspects of the approach are to argue conceptually how language can enact design. Based on those arguments, we will employ computation to give additional information basis to the critical theory. The choice of computational linguistics as the method of analysis is an approach which takes inspiration from computational language processing in order to understand the information processing behaviors of the language of design in generating meaning from language. Instead of decoding accounts of design text to understand the grammatical structure of the language of design, the computational analysis seeks to reveal how design emerges from the planes of composition that influence and enact design. Computability also adds an additional dimension of interest to the analysis. By casting language use in design in computable terms by computing over the representation of the language through its semantic and grammatical forms, the computational linguistics algorithms present language use in design as not only generating language itself (reflexively representing itself) but also as generating extralinguistic phenomena – design. Thus, we can remove the estrangement between the meaning of the language (i.e., the representation of design) and the performance of the language (i.e., the production of design).

There has been a notable absence of discussion of any other form of design representation in this line of thinking. My claim is that the line between words and non-words in design representations should not subject one to an ontological privileging over the other. But the choice of focus on language is also political. There exist people who do not sketch, do not use digital design tools, do not solve mathematical equations, or perform any of the other activities through representations one would associate with design professions. Yet, they do design; they design 'squatter dwellings' in Mumbai; they design canoes that travel vast distances over the South Pacific. Thus, what can be legitimately construed as representations of design? That design text, as a manifestation of the basic human capacity for language, can be seen to give rise to designed works de-stabilizes the principles that ascribe certain practices and activities as being a performance of design whereas others are not.

Second, in pursuing empirical information about designing by relying on designers' texts, we reinforce the connection between designer and a specific type of text – that design text is a 'special' type of text produced only by designers, using a language known only to designers, a language which is reiterating the authority of what does and does not constitute design. However, this connection assumes a free agency in the part of the designer, that a designer can willfully choose which design to perform. As with most studies of designers, the free agency to design (subject to the objectives of the design brief) is a constitutive part of design research. Designers may take in ideas from others, they may work with other designers, but only designers 'do' designing. This is how special actions and behaviors of designers become ascribed as what Nigel Cross calls "designerly ways of knowing" (Cross 2006). Important as this concept is to design professions and formal design education, in situations ranging from the design of public infrastructure projects to community-oriented development projects, what can be positively achieved is directly correlated to the exercise of designer-like behavior by non-designers.

The reconstruction of homes along the Andaman Coast of Thailand devastated by the 2004 Boxing Day Tsunami points to one illustration of this example. Not to diminish the role of architects who aided in the reconstruction by converting the words of the clients into schematic drawings for builders to follow, the exercise of the freedom of the clients to design their own housing is at least as interesting as what the trained architects (designers) did. The failure of Thai government-designed housing to satisfy the needs of the clients in comparison to the success of houses designed by the local citizens with assistance from private non-governmental organizations underscores the value of critically interrogating the aggregation of positions, the designed work accumulated from the resulting frame, and the appraisal of the frame.

These points compel us to re-think the designer with the conceptual and analytical tools for moving back and forth between what is said and written about design and the performance of design by language. In taking lessons from the post-structuralist focus on text, our focus then is to depart from the critique of the materiality of designed works to the 'goings on' of the production of the subject through text in order to depart from the baggage of anxieties induced by the subjective viewpoints and claims of what design 'is' that we would have to defend. Quoting Judith Butler, "Can language refer to materiality, or is language also the very condition under which materiality can be said to appear?" (Butler 1993, p. 31) By working within the framework of examining the constitutive role of language in design, we may gain some clarity on how design practice and a designed work become sanctioned.

My inquiry into what language does in design does not attempt to answer what design 'is'. Staking a claim to defining design in general requires some form of secondary explication. Also, I find that as soon as someone makes a claim to defining design, the definition inevitably sets up a power relationship, an inside–outside dynamic that in one way or another de-democratizes the process. As a transplant from Saigon to Los Angeles to San Francisco to Sydney from where this paragraph is being written, the latter three arguably strongholds of design in automobiles, Web, and architecture, respectively, I always feel troubled by the local discourses of design which set up an asymmetry through the mark of 'Designed by' or 'Designed in'. Especially having a historical connection to French colonialism in 'Indochina' which I continue to bear witness through readings of dialogic lamentations of French architectural grandeur in Vietnam (normally in travel brochures and history of architecture books about 'Indochina'), I find such assertions as rehearsing the historicity of a power structure which privileged the masters of design to design and others to passively receive and marvel at their work[4]. While user-centered design attracted me for a while, replacing the user at the center of a centrist-periphery model of design seemed to absolve designers of discourses of subalternation and marginalization that people have been and continued to be placed in. The repetition of urban design in the capital cities of French Indochina

---

[4] Even the strongest proponent of user-centered design, Donald A. Norman, has softened his position recently. In an article appearing in *Interactions*, Norman (2005) writes, "Yes, listen to customers, but don't always do what they say." Norman presses further on to advocate activity-based design.

and former French colonies in Africa[5] reminds us that the language of design and what constitutes design can tell us much about the colonialism of such mindsets. It seems important to note that recognition of the 'user' as not the 'center' of design but a notable vector of power in design is gaining momentum[6].

I have made the choice to deal with the question of how design is enacted through language so that the subject, the designed work, can 'speak' of its own becoming. By dealing with language, design, as a kind of being that can be interpellated as a subject, can become articulated.

This work comes on the heals of the unstoppable forces of globalization where labels on manufactured products proclaiming "Proudly Designed in Australia" or "Designed by Apple in Cupertino California" while concealing the label "Made in China" perpetuate not only an unseemly national chauvinism toward design but also tells us much about the effect of discourses about design on the business of design. By forcing ourselves as designers to be honest about the assemblage of the vectors of power that give rise to design, it may become possible for us to acknowledge that the most important thread of influence in design starts from "A line of variable direction that describes no contour and delimits no form..." (Deleuze and Guattari 1987, p. 499)

This is our roadmap. In seeking the performative aspects of the language of design through a careful examination of the language, we plan to extrapolate our findings to discover whether these performative aspects might also be candidates for the operating principles of design. And, we hope that these operating principles of design will be capacious ones which delimit design only from other human endeavors but not restrain design itself.

---

[5] On a trip to Madagascar in 2005, I was struck by the similarity, or shall I say colonial chauvinism, between the design of the Avenue de l'Independance and the Antananarivo Train Station and Nguyen Hue Boulevard and the Ho Chi Minh City People's Committee Building (Tru So Ủy Ban Nhân Dân Thành Phô' Hô Chí Minh).

[6] The New Urbanists vigorously promote the participatory design process of charrettes, intensive, collaborative meetings between designers (architects, urban planners, etc.) that seek to achieve community consensus in complex urban design projects. These charrettes, run by the Mississippi Governor's Commission on Recovery, Rebuilding and Renewal, were deployed in the re-development of Hurricane Katrina devastated communities.

2 Framing the Language of Design

It is the world of words that creates the world of things.

Jacques Lacan, *The Function and Field of Speech
and Language in Psychoanalysis*, 1977, p. 65

Design Studies

Leo Tolstoy's book *Anna Karenina* starts with the line:

Happy families are all alike; every unhappy family is unhappy in its own way.

In writing about moral codes in family and society, Tolstoy's opening line summarizes a key theme in the novel. Many factors contribute to and influence the happiness of families. The failure of any one of these factors may doom the family to tragedy. Jared Diamond cites this Anna Karenina principle in his book *Guns, Germs and Steel* to characterize the barriers encountered by societies in the domestication of animals and why the failure or lack of any of these characteristics doomed their domestication – and the concomitant doom of societies.

For designers, apparently the same principle holds true. Researchers in design studies, cognitive science, organizational behavior and strategic management (in the field of new product development) have sought to uncover the ultimate and proximate factors which influence successful ways of designing. The institutionalizing of design disciplines in universities and accreditation criteria for academic degrees in design codify what constitutes competency in design[1]. Academic journal and conference papers and popular magazine articles on designing produce a discourse about design that is at times oppositionally

---

[1] Accreditation boards often go beyond merely the accreditation of educational programs and the competencies graduates must have. They also tend to claim what design *is*. ABET, Inc. defines engineering design (design as a process) as, "Engineering design is the process of devising a system, component, or process to meet desired needs. It is a decision-making process (often iterative), in which the basic science and mathematics and engineering sciences are applied to convert resources optimally to meet a stated objective." The Royal Australian Institute of Architects (RAIA) Australian Architecture Program Accreditation and Recognition Procedure puts forth a 'Generic Statement' on design as the "exploration and reappraisal of a range of ideas and propositions that lead progressively to the eventual resolution of a coherent design proposal."

A. Dong, *The Language of Design*,
© Springer 2009

directed[2] but which nonetheless resonates the concept of design competencies. Clearly, as numerous case studies of design in industry relate, even when the designer possesses all the 'right' factors, success is not guaranteed. The artifact could be so technologically complex that not even optimal teamwork and the sheer brilliance of individuals would suffice.

So much has been written describing design that we have almost come to the point to think of design as having a set of natural characteristics. In his study of historically noted expert designers, Nigel Cross (1999b) offered a way to think about design through the nature of design. He proposed the following insights:

- Design is rhetorical. The construction of the design representation is intended to narrate a particular story that the designer wishes to communicate.
- Design is emergent. Emergent features of the design solution suggest the possible next steps.
- Design is opportunistic. It is not possible to *a priori* collect and synthesize all information about the potential design. Designers identify opportunities for potential solutions along the way.
- Design is abductive. Abductive reasoning is reasoning in which explanatory hypotheses are formed and evaluated. Abductive reasoning allows the designer to explain the layering of decisions taken each step of the process.
- Design is reflective. The designer engages in a conversation with the material (design representations) during the course of activity.
- Design is risky. There is no guarantee of success!

Lest you believe that either Nigel Cross or I believe that design is necessarily natural in the sense of an evolutionarily conserved biological program and predisposition, though I would not entirely discount these[3], I think it is important to emphasize that saying that there is no biological 'nature of design' does not mean that there is no necessity to design. The point is that what we ontologically describe as design derives from practice, from the kinds of activities that constitute design. The definition is almost tautological. As I argued in Chap. 1, the way that a designer thinks and the modes of behavior by which a designer works, while

---

[2] Basil Bernstein differentiates between institutional pedagogy, the pedagogy of design as taught in institutions and universities, and segmental pedagogy, the pedagogy of design as taught informally through popular magazines and what is said and written about design by 'informal providers'. Bernstein writes of the imbrication of these two pedagogies, "What is of interest is the interactional consequences of the relation between institutional and segmental pedagogies legitimately put together (communicate)." (Bernstein 2000, pp. 78–79) The production of discourse on design by institutions and informal providers both delimit 1) what can be considered as design, and 2) how understandings about design are to be assembled and disseminated and who may legitimately be called a 'designer'. Bernstein calls the former classification and the latter framing.

[3] The chase to find the constellation of genes associated with language is exhilarating. There was great excitement in the genetics field when a research team discovered that mutations in the *FOXP2* gene produce abnormal speech behavior (Lai et al. 2001). The popular scientific literature even called the *FOXP2* gene the "language gene". We might similarly ask if there is similar a constellation of genes for design or if linguistic competence and design competence share the same constellation of genes. We will discuss these questions in Chap. 7.

being indissociable aspects, must first be understood as 'things that happened' on the way to becoming design. I have suggested that using well-defined situations as research test-beds that then offer determinable descriptions about how the designer designs and how the designer perceives a designed work is suspect or problematic, in need of special scrutiny given the assemblages of practices which are a condition of possibility for a designer's behavior. In this chapter, we will review the main themes in design studies research and the different competences required to be acquired by a designer. Before we can undertake a project to characterize how the language of design gives expression to the becoming of design, we need to begin by understanding the current state of understanding of design and how its shortcomings give us good reason to pursue a more capacious account of design.

What I would like to provide in the next sections is a brief summary of the research that has been taken up in within a cognitive framing of design to explain design. The cognitivist framework itself is highly contentious and debated. The summaries are not intended to make corrective explanations or stake a claim as to which view is more 'correct'. The fact that debates exist as to which level of analysis (i.e., individual or social) is the most appropriate level, and which formulation of design thinking best explains designer's behavior, is rooted in, I believe, not only the various academic departments where these theories arise from but also the scale at which design is studied: an individual or a small group, by a very large, geographically and temporally dislocated group, or by a society. The point is not to get tangled up in the cognitivist-based theoretical tangles that tend to ossify the positions and debates. Rather, my intent is to question why it is that what is seen as design is strongly ascribed as a naturally occasioned practice based in mental predicates.

A Cognitive Framing

Understanding human behavior in design is both a central problem to design research and a precursor to the creation of design tools. Extensive studies of human behavior in design have been conducted within the area of design studies (Cross 1999a, 2007). Research into behavioral and social factors that influence a designer's ability to practice design successfully often study what experienced designers do. Studying experienced designers and how they approach design tasks, it is argued, offers a picture of the "natural intelligence" (Cross 1999b) experienced designers possess. So, if we're interested in looking for functional correspondences between language use and cognitive structures that facilitate cognitive processes in design, we may do well to look at how practicing, experienced designers behave.

Broadly taken, we can categorize the studies of designing along the axes of cognitive to activity-based and individual to social. Research in design cognition provides an explanatory framework for mental representations, structures, and processes in design practice. The cognitive studies treat designing as describable by a set of cognitive processes. To my mind, there are two agenda which have motivated researchers within cognitive design research. Design researchers are generally interested in the cognitive processes associated with designing as a way to explain designing and to formulate theories of design cognition. These studies researched cognitive differences between expert and novice designers (Ahmed et al. 2003) and established models of cognitive processes in design such as learning (Sim and Duffy 2004) and creativity (Cross 1997; Gero and Maher 1993). Cognitive scientists, on the other hand, approach the study of the cognitive processes associated with design to understand the nature of the cognitive processes themselves (Visser 2006). There is significant cross-over in interests between the communities and this division is only intended to account for the underlying agenda in the cited research projects.

Where the cognitive studies take cognitive behavior as the unit of analysis, the activity-based studies use an activity as the basic unit of analysis. Activity-based studies examine design processes (Goldschmidt 1992), workspace activity (Tang and Leifer 1988), and personal, practice-led 'research through design' (Pedgley 2007) to use the 'goings on' of designing as an evidence base to interpret observed practices (McDonnell 1997).

Historically, the cognitive and activity-based studies considered the designer and designer's mind or actions as centralized, independent, autonomous units operating outside of a social context. A notable exception to this view was Bucciarelli (1994) who introduced the concept of "object worlds" to describe the team environments in which designers operate, environments characterized by specialized cognitive realms determined by the designers' disciplines, professional histories and the designed work itself. Coupled with this autonomous individual perspective is that of the mind as a symbolic information processor. Models of design as search and problem solving reflect the observation that the mind performs particular cognitive processes during design. In turn, this observation is based on particular a way of thinking about the mind as a symbolic representational system. More recent research efforts based on distributed cognition move the boundary of design cognition away from the individual to include the social environment as part of a larger cognitive system. Situated cognition (Clancey 1997) corrects some failings in the symbolic information processing paradigm of artificial intelligence by accounting for the epistemology of knowledge (as symbols) as a dynamic process that develops as an agent interacts with its environment, context, and past experiences.

One of the primary issues related to this endeavor is the theoretical construct to describe the observed behaviors. Allen Newell proposed a way to structure human cognition into a "time scale of human action" (Newell 1990, p. 121) as a way to characterize cognitive levels. Using what is empirically known about how designers work, we could categorize the cognitive behavior into three bands (Fig. 2.1): individual, social and societal. Each band of behavior characterizes the extent of the social context within which designing takes place and their associated cognitivist-based models. The bands of cognitive levels provide an architecture by which to characterize the research in cognitive models of designing.

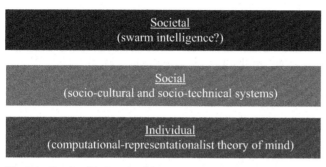

Fig. 2.1 Bands of design cognition

Given these bands of cognitive behavior, three levels of cognitive processes must be understood. The first level refers to the cognitive behavior of individual designers. This is the level of analysis that has been well researched in the literature on cognitive design research as described previously. Factors such as competency with technical design methods and tools, domain knowledge, and availability of information resources figure into individual designer's mental processes. When we turn to a consideration of the behavior of design teams and individuals' behavior within them as in the distributed cognition paradigm, second and third levels of analyses are relevant. The assumption here is that cognitive systems consisting of more than one individual have cognitive properties that differ from the cognitive properties of individuals who participate in these systems. The second level regards the cognitive system as driven by the social organization of individual and distributed cognitive processes, which are afforded by technical design tools and methods which we term their "cultural medium" (Dong 2004). The third level deals with the emergence of collective behavior in the possible absence of a shared cognitive representation of the designed work and how localized representations serve as mechanisms for self-organization. This is a nascent area of research in the field of design studies.

The commonality between these levels of analysis on designers' cognitive processes is the link between the technical tools of design and mental processes because the tools are devices that allow designers to acquire, manipulate, and realize their ideas. The commonality also suggests the inseparability of technical design tools and methods and the social context in which they are applied. The contribution of social transference of knowledge versus individual knowledge construction must be considered (McComb et al. 1999) because the tools are also devices for externalizing and communicating mental representations.

The Individual Band: Cognitivist Models

The most extensively studied band of behavior is the Individual band. At this band, the unit of analysis is the designer working essentially removed from any social

Table 2.1 Factors influencing individual designer success

Factor	Reference
Employ analogic reasoning frequently and fluently	(Leclercq and Heylighen 2002)
Employ design strategies that require reflective thinking	(Ahmed et al. 2003)
Rely on schema-driven analogizing more than case-driven analogizing	(Ball et al. 2004)
Richly link design concepts to other ideas	(Goldschmidt and Tatsa 2005)

context. The majority of protocol studies of individual designers construct experiments which remove designers from a social context. This is the level of analysis that has been well researched in the literature on "how designers think" to borrow the terminology from a book of the same name (Lawson 1997). Factors such as competency with technical design methods and tools, domain knowledge, and availability of information resources figure into a designer's mental processes. While a commonly agreed-upon cognitive model of an individual designer does not yet exist, elements of such a model have been proposed and verified. Certainly, cognitive processes and styles and their relation to design practice are well-documented.

Generally speaking, most studies into human behavior in design consider the designer working alone or the designer working as a group. In the former case, the unit of observation is the designer; in the latter, the unit is the group. There do not yet exist substantive studies delving into the changes in behavior of an individual designer in the context of the group. Taking the individual designer view, the research listed in Table 2.1 describes patterns of design thinking and cognitive structures which both influence the likelihood of success of the designer and distinguish experienced designers from novice designers. Because group design is constituted by individual designers, one would expect that the factors which influence individual designers working alone should also matter in the group design context. However, this assumption has not been tested rigorously.

The Social Band: Design as a Socio-cultural Cognitive System

At the band of the Social, the social context and the social organization of work situates the designers within a larger cognitive system. There is a wide-body of design research which describes distributed cognitive processes, how cognitive processes are mediated by artifacts, how design activity grows out of the particular situation, and the influence of the social sphere. Again, there are factors which influence and characterize successful group design, as listed in Table 2.2.

One generally accepted way to describe design as a social process is to cast design within the analytical framework of situated activity. Situated activity is a prominent analytic construct for studying and developing complex human–machine systems. In this framework, the technical and the social conditions of work operate such that mechanical efficiency and humanity are neither contradictory nor separate. Rather,

Table 2.2 Factors influencing group design success

Factor	Reference(s)
A high level of interaction with other team members resulting in an interactive and iterative combining of perspectives	(Dougherty 1992)
Share a mental model with an adequate shared understanding of the team's objectives, processes and situation	(Klimoski and Mohammed 1994)
Group comprised of balance of introverted/extroverted and judgment/perceiving Jungian typologies	(Wilde and Berberet 1995)
Have a mutual preference for working together	(Lucius and Kuhnert 1997)
Coherence of the product description diverges and converges over time, but is convergent at the end of the design process	(Song et al. 2003)
Share a common semantics of the design concept	(Dong et al. 2004)
Interfaces between collaborators, project management, and corporate strategy fluid and allow product knowledge to flow through	(Kleinsmann and Valkenburg 2008)

the concept of the socio-technical system in design (e.g., (Boujut and Tiger 2002; Lu and Cai 2001)) stresses the reciprocal interrelationship between humans and the technologies which afford their work practices.

One could express this socio-technical view on design by extending a commonly accepted descriptive definition of design as:

> The transformation of natural processes and the 'given world' through a systematic technical methodology to create an artifact that achieves a set of goals.

to

> The transformation of natural processes and the 'given world' through a systematic technical methodology mediated by social processes to create an artifact that achieves a set of goals established as a result of designers' shared understanding of the artifact's function, behavior and structure within a context defined by both the natural environment and human interests.

While the socio-technical perspective is a good start to the problem of describing design, the perspective leaves unanswered the cognitive aspects of design and the reciprocal relationship between designers and their design tools and methods. Technical design tools and methods are a "cultural medium"; they provide the structure for the transmission and propagation of cognitive states and encode patterns of behavior. Underlying the socio-technical perspective rests a layer of assumptions about how designers engage in a shared cognitive system through symbol systems and the information that they convey. The commonly accepted definition of a socio-technical system as a system composed of technical and social subsystems insufficiently captures how the social sphere regulates acceptable forms of design practice and sets up expectations as to how design is to be done.

A basic premise of the socio-cultural view is that complex performances, such as collaborative design, can best be understood as the product of a social system of interacting agents. The concept of socio-cultural systems tries to rectify the deficiencies of the socio-technical view by considering that individual and distributed cognitive processes are mediated by artifacts, that design activity grows out of the particular situation, and that the social sphere influences the situation, the choice of tools, the symbol systems and the cognitive processes of the individual designer. The basic position that the socio-cultural view aims to correct in the cognitivist paradigm is that all action takes place within a set of socio-cultural contingencies. Human action is situated in the social and cultural world; it is this world, rather than mental representations, which provide the explanation of action. Socio-cultural systems are typically described as consisting of five basic components:

1. Population
2. Culture
3. Material Products
4. Social Organization
5. Social Institutions

In describing design, the designers constitute the population and the products of their cognitive residua the material products. The culture consists of the designers' technical design tools and methods, that is, their symbol systems and the information they convey. Technical design tools and methods are a "cultural medium"; they provide the structure for the transmission and propagation of cognitive states and encode patterns of behavior. The output of design tools and methods has associated social capital, what Boujut and Tiger (2002) called the "double nature of design tools"; the tools are a medium for transferring the cognitive state of the entire design team. The social organization and social institutions within which the designer practices affect cognitive processes and development: design knowledge acquired through experience (e.g., formal education, communities of practice) informs the designer how to engage the natural and artificial world to create artifacts to satisfy human needs and desire.

This is a view I formalized in a paper describing a socio-cultural cognitive framework for design (Dong 2004). The framework connects three fundamental constructs: Suchman's (2000) situated action model which emphasizes the emergent behavior of human activity, Hutchins's (1995) distributed cognition approach which promotes the need to look at individuals, the artifacts and tools they use, and the social organization factors that influence cognition, and Wertsch's (1991) theory of mediated action which delineates the reciprocity of agents and tools.

Designing, in a socio-cultural context, could be interpreted in terms of four coupled variable groups: design stages, technical tools and methods, social processes and cognitive processes:

- Design stages: a period of time when a class of design activities occurs such as conceptual design
- Technical tools and design methods: tools (e.g., CAD) and methods (e.g., DFx) for synthesizing design artifacts; are often specific to certain design stages
- Social processes: methods and types of group interactions (e.g., negotiation, cooperation, meeting)
- Cognitive processes: mental processes at the individual and group level (e.g., exploration, selection, reflection, transactive memory, shared memory)

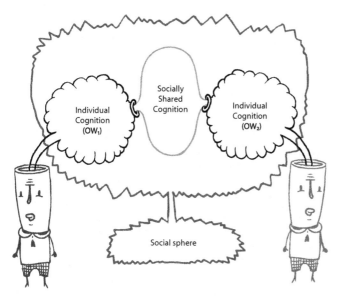

Fig. 2.2 The socio-cultural cognitive framework

The idea is to characterize design and how people design in terms of these four variable groups. Figure 2.2 illustrates the basic ideas in the framework: cognitive processes are afforded by a cultural medium comprised of technical tools and design methods; the mapping to a socially shared "object world" (OW) occurs within a social sphere.

Let's apply this framework to describe how observed cognitive outcomes are a consequence of a cognitive system in which designers coordinate their individual processing through technical tools and methods and social processes. Table 2.3(a–d) present a description of mechanical design as a socio-cultural cognitive system.

The Individual and Social bands presume the existence of mental representations of the designed artifact. That representation might be localized (the designer's mental imagery) or distributed (e.g., a shared understanding), implicit or explicit. One of the failings of the distributed cognition view is that the human mind is still, nonetheless, a cognitive system within which the processing is said to be going on in one person's head. The processing is affected by, coordinated by, influenced by, and regulated by what is going on inside another person's head and through cognitive artifacts such as tools.

The counter-balance to this strongly cognitivist view of the Social band emanates from studies promoting a social understanding of design work in which language and design tools and methods are seen as media of human action in various social settings rather than as substrates for cognitive processing. Design studies in this social perspective are not constrained by socio-technical or socio-cultural frameworks, which could be argued as bringing along a fair bit of conceptual baggage with them. Design activities, tools, and methods do not neatly fit into the socio-technical or socio-cultural frameworks when the domain of design is outside the technical disciplines such as engineering; often, the definitions for concepts such as 'culture' and 'technology' must be made so broad or vague that the frameworks lose their descriptive prescience.

Table 2.3a Socio-cultural cognitive system for mechanical design: Idea Validation Stage (Dong 2004)

Design Stage	Design Activities	Technical Tools and Methods	Social Processes
Idea Validation	Determine customer and market needs	Customer interviews, surveys, and focus groups; Ethnographic studies; Patent search; Affinity diagrams; TRIZ/ARIZ	Formation of core product development team; Assign team roles; Create "social network" (i.e., who including external suppliers should be involved)
	Determine broad product and business objectives	Product portfolio planning	Define and order product objectives (e.g., must-have features and nice-to-have features)

Cognitive Outcome: Team mental model (Badke-Schaub et al. 2007)

Table 2.3b Socio-cultural cognitive system for mechanical design: Conceptual Design Stage

Design Stage	Design Activities	Technical Tools and Methods	Social Processes
Conceptual Design	Identify essential problems	Design freedom and uncertainty	Development of team jargon and vocabulary; Decide how to handle ambiguity, uncertainty and imprecision; Distinguish specifications from constraints
	Propose function structures	Bond graphs; Design structure matrix; Mechanical design compiler	Division of design tasks among team members; Division of larger design teams into smaller teams by product sub-system
	Search for and propose solution principles	Case-based information retrieval; Method of imprecision	Consultation with experts within company, communities of practice
	Select, combine and refine into concepts	Pugh charts, pairwise comparison charts, Borda charts	
	Evaluate concepts against design criteria	QFD; Design Compatibility Analysis; Compromise Decision Support Problem	Agreement on method(s) for evaluating and selecting concepts against criteria; Establish norms for critiquing designs

Cognitive Outcome: Task and process mental model (Badke-Schaub et al. 2007)

Table 2.3c Socio-cultural cognitive system for mechanical design: Specification and Design Stage

Design Stage	Design Activities	Technical Tools and Methods	Social Processes
Specification and Design	Develop preliminary configurations	Shape annealing; evolutionary and co-evolutionary design	Establish accepted design procedures (e.g., adopt appropriate design codes); Communication and sharing of technical design issues to all team members (distributed cognition)
	Select best preliminary design(s)	Multi-objective concept selection	Reconciliation of design objective conflicts
	Refine designs and configurations and evaluate against technical and economic criteria	Design and system configuration optimization	Team learning of design trade-offs, potentials of new technologies, incorrect assumptions, etc.
	Detailed analysis of refined design(s)	Finite element modeling and analysis; Design-rule checking	Acceptance of testing methodologies, test environments, lead customers, etc.
	Review for errors, manufacturability, and cost	DFM/DFA/DFx	
	Prepare a preliminary parts list and manufacturing and assembly drawings		
	Final design analysis and verification	Finite element modeling and analysis	'Sign-off' by all team members
	Complete detailed drawings and production documents	Computer-aided drafting and visualization (e.g., solid modeling; surface modeling; rendering; animation)	Team post-mortem
Cognitive Outcome: Shared understanding of the product			

Table 2.3d Socio-cultural cognitive system for mechanical design: Prototyping and Manufacturing Stages

Design Stage	Design Activities	Technical Tools and Methods	Social Processes
Prototype production and testing	Prototype	Rapid prototype; 'desktop machining'	
	Test and evaluate	Taguchi quality, design of experiments	
Manufacturing	Fix design specifications		Transfer of design knowledge to manufacturing
Cognitive Outcome: Socially shared narrative of the overall design process			

Researchers including Larry Bucciarelli (Bucciarelli 1994), Rachel Luck (Luck 2003; Luck and McDonnell 2006), Janet McDonnell (Stumpf and McDonnell 2002), and Ben Matthews (Matthews 2007) unmoor from rationalist frameworks to reveal the complex social processes that define design. In particular, the works of Luck, McDonnell and Matthews stress the role of language, specifically in the form of conversation, and adopt a social view of language. Their social view of language, and attention to the settings within which design conversations take place, opens the design research gaze onto problems such as argumentation, negotiation, agency, power relations, and conversational strategies in relation to design activities.

For the most complex design problems, it appears as if this presumption of a shared mental representation of the designed work seems optimistic. Do designers actually have an internal representation of the designed work beyond what is necessary to complete a specific design task at a given moment in time? Can a shared understanding exist among hundreds of designers working on the design of an aircraft? Such problems have lead many researchers to question the content and richness of such representations. These questions have lead to the emergence of non-cognitive models such as swarm intelligence and social agents which negotiate representations.

The Societal Band

Designers, economists, and policy-makers generally agree that design is fundamental to the creative economy, and that the creative economy is a core generator of economic activity. Spurred by discourse on the relation between design, creativity, and urbanity (Florida 2002; Landry 2000), local governments are formulating public policy to encourage centers of creative activities to underpin design industries (Malecki 2007). Every urban region seems to want to replicate Silicon Valley, a paragon of the creative urban center. In the UK, cities, including Liverpool, Sheffield, Birmingham, Newcastle and Belfast, have 'creative quarters'. In inner suburban Brisbane, Australia lays a 'creative industries precinct'.

Modeling what makes societies creative or how it is that societies collectively design creative products is, to say the least, difficult. According to Florida (2002), policy makers just need to find the right mix of technology, tolerance, and talent. While the aim is to do so in an inclusive fashion, so that design, in the words of Gui Bonsieppe, is part of an endeavor of social justice and not "after all, a tool for domination" (2006), enacting a 'tolerant' society is a proposition which imposes a certain set of 'Western' ethics. Requiring that creative societies behave in a certain way in order to do creativity is problematic. Parceling creativity into technology, tolerance and talent is also problematic because of the unexamined assumption that people have effective opportunities to make use of technology, enjoy the freedoms of tolerance equally, or access the cultural capital that scaffolds talent. Politics aside, such economic-based models are too high-level to be useful to model how design happens on a societal level as they model outputs rather than inputs. The problem may be that modeling design at a societal level is hopeless complex.

Swarm intelligence has been proposed as one method to model design at a societal level and to handle the enormous complexity. The premise of swarm intelligence is that complex behaviors, such as design, can emerge from the succession of relatively simple actions by a collective of equal entities following simple local rules of interactions (Bonabeau et al. 1999). Models of complex behaviors arising from swarm intelligence do not imply that the agents exhibiting the behavior do not have any individual complexity or higher-order cognitive abilities. That is, to model a society of (human) designers based on swarm intelligence does not preclude the view that the humans are not complex agents. We are! Rather, the underlying assumption is the possibility to model complex behaviors as if the agents were relatively simple and interacting at a local level only. Under specific circumstances, the continuous and parallel interactions between such entities can cause a recursive process that, observed on a macroscopic level, leads to perceivable complex behaviors. Well-known examples of emergence in nature include the food foraging behavior of ant colonies, or the flocking patterns in birds and fish. Reynolds (1987) created a mathematical simulation of flocking and swarming through the movements of so-called boids, short for 'bird-objects', based on three relatively simple, iterative behavior rules to be applied on all members of a flock. Clearly, the birds have higher order behaviors to monitor their position within the flock, but it is not needed to know what those are in order to simulate flocking. Bonabeau and colleagues (1999) describe various algorithms based on the social behavior of insects to solve complex optimization problems. As yet, models of swarm intelligence to design something other than communication networks and solve combinatorial optimization problems have not been made. Clearly, this is an area worth a closer look.

A Textual Framing

The main issue, if I can grossly sum up the arguments in the cognitivist paradigm, is that, in each level of the cognitivist bands – the individual, the social, the societal – theorists posit a model of the designer as a bearer or container of a mental vocabulary upon which and with which cognitive processes take place. In the swarm intelligence model, very little is needed to be known about the cognitive processing taking place, but they are assumed to occur. The precise nature of that vocabulary and its sources and the mechanics of the processing are disputed. Yet, what is not disputed is that experienced designers seem to have certain behaviors giving them a "natural intelligence", which also make their practices (readily) ontologically describable as design. It is possibly misguided to model design as a designer who follows rules independently of the community of designers within which he/she practices, as in the individual band. It is equally misguided to model design as a designer following only the rules of the community without describing the mechanics how those rules become embedded within practice *in the first place*. It is also an entirely more complex issue to explain the mechanics of this process. In my mind, the debate within the cognitive paradigm becomes one of whether the *extent* to which the inner mental processing of designers should be (can be) analyzed by referencing only the designer and the particular occasioning situation.

Wittgenstein's concept of the language game offers an interesting way to get out of the cognitivist maze of debates. His idea of language games is that there is an internal relationship between language and activities. Wittgenstein proposed that we understand language by considering the internal relationships between the expression of language in ordinary life and what roles those expressions have in performing everyday practices. By examining the ways that language features into the specific practices important to design, the use of the language can be a part of a designer's private cognitive processing and also an element of communication, a primary function of language.

What I therefore want to argue is that if we really want to bring the understanding of design clearly into view, we must realize that what is observed as designing when it is observed through discourse or linguistic acts is the linguistic character of design. What we are witnessing is the enactment of design through language. I will use a concrete example of language enacting design to elaborate this point.

Perhaps one of the more vexing problems in cognitive design research is how designers come to construct ('have') knowledge about the artifact that they're designing. We take for the moment the social band, in which activities in design are distributed across social spaces and mediated through language-based communication. Communication is often defined as the creation of shared understanding through interaction among people. One of the purposes of communication is to establish a set of coherent ideas from which team members can develop shared understanding. From there, the stakeholders in the design process are able to articulate clearly to one another, as well as to those outside the group, their goals, purpose, design process and product. Within this framework, communication is an active process; team members do not participate passively, i.e., simply receive or transmit information. Such active communication generates new meaning for and among the team.

According to some prevailing theories of inter-personal communication (Wertsch 1991), a shared understanding between communicators is comprised of two components: a topical or contextual component and a voice component. The topical or contextual component is comprised essentially of the topic, for example; component X in assembly Y or the final colors of component X share contextual similarity through component X. Components of voice, in collaborative design, can be as simple as the jargon or specific language used in the design team. On a deeper level, voice, particularly the voice of designers operating in a team environment, becomes more defined by the ability of a designer to borrow the shared vision of a design team. In group communication, true collective understanding occurs when the team is 'on the same page' as it is popularly called. In language terms, getting 'on the same page' requires achieving some group acceptance of a common set of vocabulary and a common grammar. This is particularly important when designers on the team come from different disciplines or backgrounds. Similarities in voice between designers in a group are critical to progressing through the design process.

Perhaps more important is the concept of a collective voice. Bakhtin wrote extensively that effective group communication occurs when the group shares a voice, and in fact the intra-member dialogues are merely extensions of this shared voice (Wertsch 1991). When a speaker wishes to address a group or another speaker, the speaker borrows a voice from the collective group. This collective group voice is dynamic and changes as the communicative requirements progress, but effective group communication occurs when all members are able to borrow from and relate to this combined group voice.

It is unclear whether the common topic and voice signifies a shared mental representation or if the manifestation of the common topic and voice is purely linguistic. Research in information use in design (Baird et al. 2000; Lloyd 2000) have surfaced that words and phrases used by designers in the design process often capture personal experience and contribute to a wider narrative at the team, project or corporation level. There is no suggestion, though, that all speakers share the same mental representation of the narrative. What is known is that the interaction between the speaker, the speaker's intentions, and the voices that subtend the speaker shape the narrative. Processes of a communicative nature produce a "speaking personality" in any social setting (Wertsch 1991). Individual cognitive processes will be conceptually 'close' to social communicative processes because the process of understanding the communication between speakers demands that each speaker understands not only the vocabulary of the communication but also the corresponding context in which the communication takes place. This does not mean, however, that having a shared voice is the same as having a shared mental representation. Although topical and voice similarity may not indicate a mental shared understanding, their linguistic realization is at least sufficient evidence of the role of language in forming a shared voice between the communicators.

The analytic approach of viewing language as having social communicative aims requires us to consider why certain forms of language were invoked. Focusing on internal cognitive processes preceding the communication privileges too much an internal, 'willful' agency in producing the language. I agree with the theory of mediated action that the purpose of the communication is to establish a set of ideas, for the speaker to 'get a thought across' in a way that is acceptable to the audience. Their picture of language mediating internal cognitive processes though is incomplete in so far as their perspective fails to address the role of language producing new realities rather than merely relaying existing experiences. My point of departure is that language is also a way to produce a reality. As opposed to a merely communicative purpose, namely the transmission and reception of experiences, language serves an active role of generating new realities. Embedded in the language of design is information regarding the design concept, a concept the designer is attempting to communicate and advance.

Language is enacting the design concept as it is representing and manifesting individual cognitive processes, as it is a cognitive artifact supporting those processes. I insist that language use in design is not something we should use to shore up the cognitivist paradigm, as a convenient means to account for cognitive processing for example. The issue is not the link between cognitive processes in design and language, i.e., is the production of language part and parcel with basic cognitive operation such as mental modeling and framing. It is neither the issue to categorize which circumstances of language use in design are appropriate for us to regard language as mirroring the cognitive processes of designers. Instead, language use in design has to be explained with reference to the reality-producing nature of design and the ways in which discourse about design become internalized such that what when we read the discourse, the language that is enacting design seems entirely 'natural'.

Through Wittgenstein's language games concept, we have a template for thinking about the role of language in design. Language in other words, should not be taken for granted. Equally, it should not be superficially denied as being as closely related to design as is any other semiotic activity or comfortably dismissed as something that is a necessity for design as much as it is a necessity for social living. By recognizing language in terms of its functioning for enacting design, design cognition at an individual level or at a social level are not inevitable corollaries. In my view, it can be said that language use in design *is* the designing and not an approximation to the 'reality' of designing.

Accounts of Design

I interpret these design studies, as this field has come to be known, as inquiring into the conditions from which designing emerges. If language is part of the matrix of conditions from which designing and the designed work actualize, as our thesis conjectures, what we need to do is to consider how the language of design, in accounting for itself[4], that is describing and characterizing design, has *design* as an effect. What I am suggesting is that an account of design through language is one way of decontextualizing design from technical and cognitive frameworks, as discussed above. Design emerges as a subject within its own account, an account which can internalize these preceding frameworks but which is not contingent upon them. This perspective stands in contrast to design studies which have set up the following frameworks by which an account of design can be understood:

- Design as a both an individual cognitive activity and a socio-cultural cognitive system
- Design as the construction of knowledge representations through knowledge production and integration
- Design as a reflective practice in which the designer's situated reasoning and emotions engage in a conversation with the design representations

---

[4] I borrow this terminology from Judith Butler's book *Giving an Account of Oneself.*

They have set up the major themes for making sense of design and orchestrating the ways that an account of design could be presented. Open issues remain, of course, including cross-cultural variations (if any) in cognitive processing in design, and the acquisition of the cognitive processes. What we want to do in our project though is to consider the limit of such constraints. If we choose to understand design from technical, social and cognitive frameworks, what do we give up? If cognition, social science, and technical positivism are the starting points from which language use in design could be framed, then we seem to limit our understanding of design by externally operating on the subject – design. The specific theoretical models within these frameworks often conflict, forcing design researchers to re-visit ideas anew. New cognitive models emerge, which are then applied to understanding design, often with a 'lag' function. As a consequence, design studies research can sometimes appear to be on the 'back foot' awaiting for advances in related fields in order to allow advancements in its field of research.

Let us take a slight diversion into psychotherapy. In psychotherapy, the patient, with the assistance of the therapist, attempts to reconstruct or recollect past memories as a means to make sense of what the patient has lived through. This narrative reconstruction of the patient's life is used for purposes such as helping the patient to cope with past and current problems, to identify the sources or root causes of these problems, and to develop tools with which to handle new problems as they arise. One of the roles of the therapist is to intervene in the narrative reconstruction to help the patient to 'put the pieces of the puzzle' together, as it were, in order to establish a coherent and complete account of episodes in the patient's life or accounts of the patient as a subject. The therapist then attempts to formulate explanations for underlying themes or recurrent events in the accounts given by the patient in order to prescribe courses of therapy for the patient. The unexpected similarity between psychotherapy and design studies is that design researchers are the 'therapists', designers are the 'patients', and the designers' accounts of design through language are what design researchers could use to provide explanatory frameworks about design. Instead of using psychology to fill in the gaps in the patients' accounts and then to explain them, design researchers currently use cognitive science or social science to establish conceptual models of design which are then 'discovered' and validated through accounts of design. One problem with this approach is the unknowing of why language in the accounts behaved in particular ways.

To address this problem, we need to discuss the language of design as descriptive framework which has dealt with this very issue. Donald Schön defined language use in design as "elements of the language of designing" (1983, p. 95), that is, as descriptors of what takes place during design and of what consequences actions have as described by the language. Reading language use in design as representing design is taken as a given, and certainly taking the position that language is a representation of cognitive processing, incomplete as it may be, is not without its merits or useful outcomes.

This representational notion is strongly linked to the idea that the mind is an information processor and information processing theories of language. If we think of the human mind as essentially a representational system, as Margaret Boden (1988) explained, then design is realized by the construction, organization, interpretation and transformation of mental representations. The linguistic residue represents the outcome of the cognitive processes. They constitute the structure and outcome of cognitive processing in design.

The basic assumption of an information processing model of the mind is that symbolic communication, such as language, portrays the cognitive processes of the designers. Certain types of knowledge are stored as semantic memory and, possibly, distributed throughout the design team. Semantic and grammatical structures of language-based communication encode the designers' thoughts. The consequence of such a model is to see all language use in design as occurring through a prescribed cognitive structure. We find these prescribed cognitive structures implied by researchers who base their work on symbolic artificial intelligence. Through a formal functional requirements language expressed as a pair of transitive verbs and nouns, Jacobsen et al. (1991) attempted to capture the functional-structural reasoning processes engineering designers conduct. Even where words were annotations to drawings, such as the body of work in design rationale capture and re-use systems (Garcia and Howard 1992; Regli et al. 2000) and case-based information retrieval systems (Wood III and Agogino 1996), words were treated as passive indices of the mental structure of the designers' minds. In all of these approaches, language-based design representations could be conceived as linguistic behaviors that represent the structure of thought in the design process.

The connectionist models of the mind take some corrective measures to the symbolic reasoning view. One of the basic tenets of connectionist models as a model of the mind, in contrast to symbolic reasoning systems, is that the mind could be described in terms of a network of interconnected units rather than as a symbol processor. Concepts emerge from connections across schemas (Coyne et al. 1993) rather than through a prescribed reasoning process (Benami and Jin 2002). A consequence of the connectionist view is to regard the mind not as a symbolic processor but rather as comprised of individual units in states of connectedness. Despite these corrective measures, it is perhaps untenable to suggest that humans are not symbolic creatures whose behaviors are not formed and enabled by symbolic reasoning, which we are (Deacon 1997). Thus, the connectionist model, while useful as a computational model of the mind, does not seem a plausible explanation for framing why language behaves in particular ways.

Seeing the language of design as descriptors, though, means we give up the potential to see words as a model of future, possible realities. We end up accounting for language's origins in the mind rather than its reality-producing possibilities. We intuitively know that we do not use language just to represent an idea. Human thinking and ability to develop knowledge is heavily constrained by and made possible through language. Language is essential for the production of knowledge. So, language must also been seen as playing a constructive role in design. The use of language in design is effective in contributing to designing itself. Presenting designers with lexical concepts semantically related to verbalizations can spur new ways of thinking about a designed work (de Vries et al. 2005). Language bridges relations among distributed knowledge (Dong 2006), possibly serving as the conscious interface between inter-related cognitive systems. There are surface linguistic features correlated to creative design outcomes (Mabogunje and Leifer 1997). Asking 'generative design questions' during design can facilitate types of cognitive processes such as divergent thinking (Eris 2003). None of these observed linguistic behaviors and effects of language have anything to do with describing design. If anything, language is mediating design activities.

The psychologist Vygotsky would not find design studies' findings on language's mediating role in design surprising. In one intriguing study, Vygotsky compared the way that literate and nonliterate people reasoned about objects. In the study, the subjects were asked to group together the objects 'hammer', 'saw', 'hatchet', and 'log'. The literate subjects grouped the objects 'hammer', 'saw', and 'hatchet' based on the abstract meaning 'tool' because these objects are all types of tools. The illiterate subjects, however, grouped the objects 'hammer', 'saw' and 'log' since these objects would normally be associated with the action of wood-cutting. That is, Vygotsky found that the literate subjects reasoned using linguistic categories and linguistically created realities whereas the non-literate subjects reasoned based on practical experience (Wertsch 1985, pp. 34–35). The effect of language in design thinking plays within a set of complex roles, influencing not just the ways that designers think and express their thinking but perhaps regulating the way that they approach designing. So much, then, for the language of design as only descriptive.

From our point of view, the interesting question is in what ways language performs design. So, when cognitive design research appeals to the theory of mediated action to suggest that language operates as an agent for mediated action, what should really be emphasized is that language is a 'tool' for making new linguistic realities. Language achieves various reality-producing outcomes: it projects possibilities (e.g., through modal verbs, verbal auxiliaries, interrogatives); it forms design concepts that only exist linguistically; it negotiates the value of design concepts and ends up saying to us, 'Yes, this is the concept to proceed with.'

The research literature on design studies is effused with terms such as mental models, shared understanding, teams, groups, groupthink, distributed cognition, group cognition, and communities of practice as various ways to explain variations in and to account for designing. While I am slightly overstating the case, language is given short shrift. Certainly, design research has not made the type of cultural theorist questioning about how the production of text is related to the production of the subject, the designed artifact. Language use in design is not neutral; it cannot be regarded as an artifact of symbolic processing without acknowledging its instrumental role. Its instrumental role cannot be argued without making the case how language came to have regulatory and influencing functions on what could be designed. Any account of design practice and the designed work through language implicates language with working to actualize design.

If language use in design plays constitutive, representational, constructive and instrumental roles, then we need a broader and more capacious account of the interconnections and complementarity of these roles. However, the theorization of the relation between language and design is notoriously difficult, traversing as it does a complex terrain which covers both competing viewpoints on design and competing viewpoints on language. The concept of design has no 'natural' grounding; design has its ontological basis in praxis. Language, conversely, has biological, social and cultural roots, with a wide array of analytical formalisms and paradigms by which to analyze language use. How precisely do we describe linguistic processes so that the mechanisms of design through and in language are made explicit? And, do we gaze from the perspective of language and linguistic theories and how those perspectives inflect upon our understandings of design, or do we start from a perspective on design and then seek out linguistic theories congruent with them? Each perspective from which to frame the language of design has its own set of consequences and partial answers.

This book, and the theory it proposes, aims to conceive of a linguistically-informed conceptual apparatus to examine the relationship between performed design activities and what is said and written during their performance. When accounts of design are realizing design, these accounts can provide an analysis of the conditions under which design is constituted in language. This analytical stance makes it more possible for design to tell a truth about itself without reliance on external conceptual frameworks that we seek to validate through these accounts. The aim is to think conceptually about language itself.

Judith Butler offers the following insight into the problem of receiving an account of a subject when the terms of reference are set up externally:

> These terms are outside the subject to some degree, but they are also presented as the available norms through which self-recognition can take place, so that what I can "be", quite literally, is constrained in advance by a regime of truth that decides what will and will not be a recognizable form of being. (Butler 2005, p. 22)

If, instead, we turn to an understanding of design based on its own accounts of itself, then perhaps we set a stage for design, through accounts in language, to craft itself. To say this rather playfully, I am suggesting that we let design's own discourse design *design* itself. As Jacques Lacan stated in the epigraph, the question we need to grapple with is how words produce realities. This will require that our language theories and computational tools pay attention to actions rather than just representation. To do so, we appeal to the conceptual, reality-producing effects of language as the framing for the language of design. The question I would like to turn to then is a consideration of how describing design also produces design, in which the proscenium of performance is the account of design in design texts.

The Performativity of the Language of Design

Our focus on language doing something and producing a reality (the design process and the designed work) is closely linked to concepts on language producing that which it represents (performativity) and the production of the subject in discourse (psychoanalysis). Language probably shapes reality in many ways, as its role in shaping reality is theorized by many philosophers and empirically examined by linguists. As our focus will be on developing an account of language producing design, I shall turn first to a broader and more philosophical view on language and its role in shaping reality and how these ideas support my view on how language is constitutively involved in producing design. That language does design rather than merely represents design is fundamentally the concept of performativity.

The concept of performativity traces its origin to the speech act theory concept of performative utterances, first described by J.L. Austin (1962). When the production of an utterance *is* the performance of an action, Austin called this type of utterance performative. The Austinian formulation of performativity states that utterances produce through naming. In a performative speech act, the aim of the speaker is not to describe what is being 'done' but to 'do it'. He uses the examples of marriage, 'I do', and naming a ship, 'I name this ship the *Queen Elizabeth*', to illustrate speech acts which 'indulge' to 'do' rather than state facts (Austin 1962, p. 6). The equivalence to design is that the realization of lexicalized concepts is a form of design practice; in explicitly stating the designed work linguistically, the language produces the designed work. Language is 'doing' the designing. This fundamental idea of an instance of language actually producing something new in reality is a key philosophical point.

However, the language of design cannot produce design through the grammatical form with "all verbs in the first person singular present indicative active" (Austin 1962, p. 56), the grammatical form of performative utterances Austin formulated. "I declare this a widget" is not believable to register a widget or how the widget became one – that is, how it was designed. In fact, Austin points out that performatives may be 'felicitous' or 'infelicitous', that is, performative utterances may succeed or fail to produce the reality stated. Austin establishes in-depth criteria for 'felicitous' and 'infelicitous' performative utterances such as "accepted conventional procedure" (Austin 1962, p. 26) and a social structure to execute the performative (Austin 1962, p. 36). Along with his concern on 'felicity', Austin states that performative utterances are indeterminate as to the truthful expression of realities and the truthfulness of the "inward and spiritual act" by the speaker (Austin 1962, p. 10). Instead, he states that performatives, at least, evoke a promise of action, if not the action itself (Austin 1962, p. 11). Austin's concern with the explicit grammatical form of performatives may have hindered him from theorizing more on how language produces a reality. He rewrites performative utterances into various grammatical forms other than the first person indicative active to illustrate their potential linguistic realizations. It may also account for his partial dismantling of the concept (Austin 1962, p. 52 and p. 145). Yet, his template for thinking about words producing a reality, and especially his theorizing on the matrix of conventions and social structures that bring 'felicity' to performatives and the 'inward' intention of the speaker versus the outward 'truth' of the speech act, now extends beyond speech act theory. His concept of 'felicity' is especially cogent to design theory. The potential to map the felicity of performatives to the completion or incompletion of design activities would be invaluable.

The philosophy of performativity has been eloquently rethought by philosophers in gender studies and feminist theories on the representation and production of the body. As a means to question how the performative derives its force to produce reality, Eve Kosofsky Sedgwick (2003) uses the term 'periperformative' to describe the scene around performatives. Sedgwick contends that periperformatives sanction (or circumvent) the power of the performative to produce an effect. She illustrates the concept of the periperformatives around the performative utterance 'I do'. The saying of 'I do' to marry a couple derives its power to create the reality of a married couple in part by the surrounding elements including a justice of the peace, the witnesses, and the marriage certificate. The lack of these periperformative elements dooms the performative. A linguistic consequence of her point is the futility of identifying specific textual passages (such as 'I do') that signify the performative. Instead, we may do well to pay attention to the overall patterns of the text and interstices of the semantics and grammatical structures of the language of design.

This ground is propelled most forcefully by Judith Butler's widely-cited performative theory of gender which proposes that gender is not biologically ascribed to sex; rather, gender is performatively produced. In her book *Bodies That Matter* (1993), Butler interrogates the production of gender through a careful interweaving of psychoanalysis, phenomenology, and, particularly, speech act theory. Butler takes speech act theory further by explaining how social systems of meaning precede the subject and that, paradoxically, embodying those systems in our actions make them appear natural and normal. However, Butler is careful to note that the choice of which gender to perform is not necessarily a willful choice. Butler relates the performative production of gender to discourse in the following passage:

> For discourse to materialize a set of effects, "discourse" itself must be understood as complex and convergent chains in which "effects" are vectors of power. In this sense, what is constituted in discourse is not fixed in or by discourse, but becomes the condition and occasion for further action. This does not mean that any action is possible on the basis of a discursive effect. ... The power of discourse to materialize its effects is thus consonant with the power of discourse to circumscribe the domain of intelligibility. Hence, the reading of "performativity" as willful and arbitrary choice misses the point that the historicity of discourse and, in particular, the historicity of norms ... constitute the power of discourse to enact what it names. (Butler 1993, p. 187)

Butler emphasizes that performativity should not be thought solely as "the act by which a subject brings into being what she/he names, but rather, as that reiterative power of discourse to produce the phenomena that it regulates and constrains." (1993, p. 2) That is, Butler suggests that citation and temporality are lodged in the realization of performativity. Whereas Sedgwick questions the scene around which performativity gains its power to produce an effect, Butler contends that the 'successful' production of gender is an act based on a system of meanings which gained their authority to produce an effect through citation and the historicity of norms.

In summary, there are three main aspects to performativity that relate to our thesis: that language produces an effect, that language has the power to produce the effect, and that language has the authority to produce the effect. Understanding the performativity of the language of design entails:

1. Recognizing that the language of design performatively produces reality. However, it will not produce reality through the grammatical form of first person present indicative of Austinian performatives. Of frustration to design researchers and linguists alike, neither Sedgwick nor Butler gives us explicit guidance on how to locate the performative semantic and grammatical aspects of language. If performatives do not necessarily exist as first person present indicatives, how do we empirically locate them and distinguish between felicitous and infelicitous performatives? Distinguishing between felicitous and infelicitous performative language would be equivalent to identifying instances of successful and unsuccessful design practice, which is of value to design research. Computational linguistic-based research on language use in design will confirm what Sedgwick speculated, that we should not expect to find (nor do we actually find) the performative aspects of the language of design explicitly inscribed in one or two sentences or paragraphs. Instead, we (will) find them in the patterns of language use.
2. Recognizing that language use in design re-iterates and re-confirms the authority of a community of designers to delimit what is considered to be design practice and what is a recognizable designed work through reiterative textual performances. Austin alluded to this when he stated that felicitous performatives require conventions of beliefs which support people to place trust in the intentionality and the truth value of the performative utterance. In order for language to performatively enact design, what the language 'says' must accord with an entire set of preceding discourses which determine what content must exist in order for the account to be a believable one of design rather than something else. It must support the reader to believe, 'Yes, I see design happening in this account.' This assertion relates primarily though to Butler's critique of the hegemonic conventions and ideologies that become incorporated into gender performances. The language of design itself can offer a site of critique on the processes and norms which become inscribed into design. The premise is that the semantic and grammatical structuring of language use in design is also a message that carries with it 'codes' which reiterate systems of meanings about design/ing. This point raises the question of how one analyzes what other messages the language carries and how they can be registered.

We are now in a position to state our theory on the performativity of the language of design. In the prior chapter, I offered a baseline theory on the language of design, one which dealt only with its representational characteristics. I propose to extend this theory in light of the linguistic and cultural studies theories on performativity and the field of design studies. I propose that the connections between the language of design and the realization of design are best explained by the concept of *performativity*.

Theory on the Performativity of the Language of Design

Premise – The structural units of any language of design consist of a set of symbols, a set of relations between the symbols, and features that key the expressiveness of symbols.

Theory – The language of design performatively enacts design through its semantics and grammatical structures.

The theory stakes four claims:

1. The language of design enacts design through three performative operators: 1) aggregation – to frame lexicalized concepts that determine what will qualify as the material of design; 2) accumulation – to connect lexicalized concepts to achieve a transformation of words into a materiality; and 3) appraisal – to give affect to concepts as a means to co-construct and shape subjectivity.
2. The performative operators of the language of design locate themselves in the overall terminological patterns of design text and the interstices of their semantics and grammatical structures.
3. The 'felicity' of the performative operators map to 'successful' design outcomes.
4. The performative operators register how designers and design disciplines negotiate authority.

Through these performative aspects, the language of design enacts design and realizes the designed work. Design text is a receptacle for the language of design. Design text draws upon assemblages of the language of design to produce the designed work. Design text is intertwined in the ontological circuit of recognition by channeling the performative aspects of the language of design toward enacting design. In harnessing and representing that which can be conversed and said, the text enacts design through the performative operators, thereby producing the designed work.

We now devote our energies towards dealing with each of these claims. The following three chapters will delve into the first three claims. The subsequent chapter examines how the language of design negotiates and registers authority in various fields of design. As I alluded to in the prior chapter, computational linguistics will figure prominently as the mode by which I link the language of design to the enactment of design. It is one thing to say that the language of design is performative; it is quite another to show the performative linguistic patterns and reveal their mechanics. This is the purpose of computational linguistics. Before proceeding onward, let me attempt to explain why computational linguistics as a set of methods to understand what is coded by a natural language are useful apparatus serving the needs of shoring up the theory of the performativity of the language of design.

Why Computational Linguistics

For quite some time, the predominant approach to understanding design is by understanding human behavior in design. Researchers in the field of design science rely on social science and cognitive science methods such as protocol analysis, ethnography and surveys to instrument laboratory-scale experiments and field-scale case studies. Researchers have developed specialized coding schemes to translate what is said during designing into data about cognitive processing in design. The qualitative research method of verbal protocol analysis (Ericsson and Simon 1993) has been well-established as a tool for studying the cognitive processes of designers. These research methods have track records of significant theoretical and analytical value towards deriving rules and representations of designers' minds both from a representational theory of mind to situated activity. For design teams, where cognitive activities are distributed across social spaces, the method of verbal protocol analysis tends to suffer under pragmatic problems which limit the scalability of the method. The scale and complexity of contemporary design projects such as the venues for the Beijing 2008 Olympics and new commercial aircraft including the Airbus A380 and the Boeing 787 question whether the findings of cognitive design research *can* adequately account for how design practice is produced at this scale. While it may be possible to scale-up tools and methodologies from linguistics, cognitive science and social science to study design, it is not *a priori* obvious how this could be done.

To progress towards understanding the type of design that is practiced in industry, on the scale of projects such as those described above, we need to move beyond standard methods in linguistics and cognitive science such as text analysis, discourse analysis and protocol analysis. The matter of scale-up aside, here are two key reasons why I believe that manual analysis of design text may not uncover the evidence needed:

1. Design is not packaged into clauses. It is a phenomenon arising from the corpus.
2. It is the combination of the effects of aggregation, accumulation and appraisal that enables us to relate the performative operators to the successful/unsuccessful performance of design rather than the existence of individual performative clauses.

Computation will be integral to our study of the performative operators of the language of design because of its capacity to take a corpus-level view. In fact, computational linguistic algorithms generally perform better at handling corpus level phenomena than clause level phenomena. One of the reasons why we eschew manual analysis of language use in design to uncover its performative operators is not just the labor associated with the analysis or even the scale-up problem. The aim is to generate a computationally-derived view of how the language of design enacts design based on the way that the language is assembled. Specifically, we will apply the machinery of computational linguistics.

The field of computational linguistics researches computational models of linguistic phenomena. The maturity of the field has opened new opportunities for a study of the language of design by asking for computational interpretations of critical theory concepts. While computational linguistics is far away from full natural language understanding, systems such as Google have proven far more utilitarian than a talking computer. It is generally acknowledged that a complete natural language understanding system which can recognize human speech, understand natural language, decide upon external communication (a response), generate the response, and then output the response as synthesized speech is not entirely feasible. Nonetheless, numerous successful natural language systems exist which can understand finite-state human speech, ascertain sentiment in written text, and grade student essays, among many examples. A natural language system as complete as HAL in the Stanley Kubrick's *2001: A Space Odyssey* may never exist. Yet, HAL-like systems already allow us to order a taxi or a pizza by talking to a voice-recognition system. Although the systems tend to be trained for local accents, the reliability of the systems is no less than remarkable. Computational linguistics has a potential to deal with the scale and complexity problems that cognitive and social science cannot.

Computational linguistics as a research tool in design studies, while increasingly gaining popularity and acceptance, has not attained the same level of implementation as in fields not directly related to studying language. In computational social systems, text processing has been used to study communication patterns in social groups and to model the content of public conversations (Sack 2000). Research in medical informatics relied on text processing is an integral tool for the formation of the Unified Medical Language System. The computational linguistic technique of latent semantic analysis (Landauer et al. 1998) has been applied to analyze the knowledge representation through language-based communication (Landauer and Dumais 1997). The psychologists Pennebaker and King (1999) developed the Linguistic Inquiry and Word Count (LIWC) software tool to study large bodies of written text by patients in order to examine how the words people use might also express their mental and physical condition. Some important elements of these studies include the abstract use of language, language as shared social ties, and language as an active constituent in creating notions of 'self'. These elements will figure into the computational linguistic algorithms presented to illustrate the performative aspects of the language of design.

Computational linguistics is an apt method of analysis to uncover concepts about how language enacts design owing to the information processing capabilities of computation. Computation necessarily implies an 'output' – design. When taken together, the methods of computational linguistics presented in the next three chapters, all of which came from applied research in information retrieval and text analysis, explain the poorly understood phenomena of performativity. Moreover, computation's repeatable techniques for handling word meaning and syntax as nothing more than symbols means that other theories that postulate performative principles in other representations (of design) could be effectively isomorphic to these computational models, given an alignment between the features of the symbol systems.

To an extent, the theory of performativity of the language of design came as a result of the observed empirical successes of the computational linguistic algorithms in explaining aspects of design. However, the computational text analysis algorithms that we deploy to investigate how language enacts design should not be seen as direct maps, or even objective 'proofs', of concepts from linguistics and gender theory. My disposition is to take ideas and methods that form one academic field and place them on the Lacanian mirror-stage (Lacan 1977a), as it were, to see which ideas and methods inflect from an Other academic field. "[T]he signifier requires another locus – the locus of the Other, the other witness, the witness Other than any of the partners." (Lacan 1977b, p. 305) One such inflection that takes place when concepts from linguistics and gender theory are placed in front of the mirror-stage is the appearance of the Other, computational science and the associated machinery of computational linguistics. With this book, I hope to offer an example of an attempt to practice computational science with a critical theory inflection, to reduce the dissonance between fields which have said much about design.

What is important here is that the approach is based on thinking about language in conceptual and pragmatic ways as 'first principles' for thinking about design. The motivation is to work through the interactions between critical theory and computational science on the one hand and language and design on the other. The inflection of computational science and its premise of the gravity of empirical evidence with critical theory and its the currency of philosophical arguments which are on balance more 'right' sets a stage for turning linguistic information into design knowledge. Design is now envisioned at the scale of lexis, wherein words in the language of design assemble into an ever-shifting concept of design.

3 Aggregation

> *[A pattern language] says that when you build a thing you
> cannot merely build that thing in isolation, but must also
> repair the world around it, and within it, so that the larger
> world at that one place becomes more coherent, and more
> whole; and the thing which you make takes its place in the
> web of nature, as you make it.*

Christopher Alexander et al., *A Pattern Language*, 1977, p. xiii

Form Words to Concept

If you have ever listened to or given a design presentation, you will already know that the favorite question to designers is, "But … what was the concept?"

What is this design concept that we intently listen for? This is a question that design tutors frequently ask from their students and what practicing design professionals try to communicate to their clients. But, asking a designer to define concept (What is a concept?) is like asking a designer to define design. There is a restless nature to design concepts. The concept of a concept is dissolved in the designers' activities. Most tutors tell their students that they 'know' a design concept when they 'see' one, and sometimes believe that a concept is crucial and other times not. Moreover, a tutor might sometimes forget that the tutor acknowledged the existence of a concept in the student's work just last week. Like the tutor, what were we listening to in the words that allowed us to recognize that a concept had been presented to us? To answer this question, we must become interested in not only where the concept was located in the words (to use a spatial analogy) but also how the words produced the concept.

The quote by Christopher Alexander in the epigraph seems to offer us a partial answer. What Christopher Alexander writes with regard to a pattern language couches a designed work, the effect of a design concept, in terms of what the 'thing' can do – make a place more 'coherent'. Even if we cannot agree what a concept could normatively be defined as or agree upon what a concept is when presented with one, we can probably agree that the formation of the concept involves the alignment of a complex, multitude of ideas that a designer can draw upon to achieve coherence. The composition of the designed work with its context must make a coherent whole. If the goal of talking about and writing about a design concept is the formation of the concept itself, we need to think about what language does in this process of framing the complex. What we need to explore in this chapter is how the language of design functions to assemble the potential constituents for the expression of the meaning of the concept into a coherent frame.

A. Dong, *The Language of Design*,
© Springer 2009

Charles Jencks maps out various ways that language is deployed to communicate an architectural concept. In the chapter titled "The Modes of Architectural Communication" in his book *The Language of Post-Modern Architecture* (1981), he lists metaphor, words, syntax, and semantics as parts of the repertoire of architectural prose. The architectural prose for Jencks is part of the lexis for symbolizing elements of architectural works and for producing meaning about the works. Through the repertoire, architects produce visual codes, a visual mental imagery. Jencks is careful to point out however, "a crucial difference between the 'words' of architecture and of speech" for which "architectural words are more elastic and polymorphous than those of spoken or written language, and are more based on their physical context and the code of the viewer for their specific sense." Architectural communication is thus a practice of the lexical articulation of meaning in the field of the built environment. Jencks comments that producing the concept requires the architect to "make use of the language of the local culture, otherwise his message falls on deaf ears, or is distorted to fit this local language." (Jencks 1981, p. 37)

An aim of the account of a design concept is to establish meanings that constitute its fundamental thematic origins. The design concept is explicitly and implicitly inscribed into language such that the language begins to structure the mode of the designed work's realization. While the concept may come to the designer enigmatically, in which its inscription in language may appear opaque and incoherent, there is something in the language that survives its potentially inscrutable narrative. The articulation constructs the design concept but also reconstructs and recovers prior patterns and meanings that served as the basis of the design concept's formation. To do so, the language intersects a multitude of knowledge and experiences. The text draws from a wide body of knowledge and experiences to establish the meaning potential that is assembled into a frame for the design concept.

To illustrate what we shall name as the concept of aggregation, I will draw from the scenario Schön described of Quist, the studio master, and Petra, the student, using words to bring each other toward "congruence of meaning" (Schön 1983, p. 81) in the production of a design concept[1]. It seems important here to consider more carefully what the words are doing in bringing Quist and Petra toward a "congruence of meaning".

Here, Quist assists Petra in dealing with the problem of siting the building. In the dialogue between Quist and Petra, their communicative acts display a realized expectation of agreement on the structuring of language that Petra can employ to help Quist constructively interpret the concept. The concept for the site is described in terms of the building's relation to other objects and the mutual orientations produced rather than, for example, energy consumption due to orientation. Evidence that is not interpretable or are irrelevant or that fall outside of the style of design communication of the architectural tradition that Quist models for Petra serve as constraints on her communicative acts. When Petra responds with a technicaliza-

---

[1] Throughout the book, I will make reference to familiar texts of language use in design to illustrate my linguistic interpretations. I encourage you to (re)visit the original texts and interpretations to contrast both the style of analysis and the concepts that are central to the inquiry.

Table 3.1 Conversation between Quist and Petra (Schön 1983, pp. 82–85). Reprinted by permission of BASIC BOOKS, a member of Perseus Books Group

Speaker	Segment	Statement
Petra	1	I am having trouble getting past this diagrammatic phase – I've written down the problems on this list. I've tried to butt the shape of the building into the contours of the land there – but the shape doesn't fit into the slope. I chose the site because it would relate to the field there but the approach is here. So I decided the gym must be here – so I have the layout like this.
Quist	2	What other big problems?
Petra	3	I had six of these classroom units, but they were too small in scale to do much with. So I changed them to this much more significant layout (the L shapes). It relates one to two, three to four, and five to six grades, which is more what I wanted to do educationally anyway. What I have here is a space in here which is more of a home base. I'll have an outside/outside which can be used and an outside/inside which can be used – then that opens into your resource library/language thing.
Quist	4	This is to scale?
Petra	5	Yes.
Quist	6	Okay, say we have introduced scale. But in the new setup, what about north-south?
Petra	7	This is the road coming in here, and I figured the turning circle would be somewhere here.
Quist	8	Now this would allow you one private orientation from here and it would generate geometry in this direction. It would be a parallel …
Petra	9	Yes, I'd thought of twenty feet …
Quist	10	You should begin with a discipline, even if it is arbitrary … The principle is that you work simultaneously from the unit and from the total and then go in cycles …

tion of the orientation, Quist does not affirm her technicalization, but, instead, comments that Petra needs a set of working principles. In setting this principle, he is effectively borrowing from the voices of prior architects. The effect of cognitive activities in design in transforming representations may be seen acting here as transforming other architects' voices by Quist and Petra intentionally selecting from among them in such a way as to communicate the design concept.

The semantic and grammatical structure of their dialogue follows a set of expectations in accordance with not only what is needed to explain an architectural design concept, but also what is contained within the language to explain the concept. For example, she selects empathy to explain for the rationale behind the L-shape layout by stating that the layout "is more what I wanted to do educationally anyway". In making public evidence to support her concept, Petra looks for evidence in past thoughts and must decide which of these thoughts fit together, which of these thoughts could function as a display mechanism that are recognizable and interpretable to make words into a framing of design concept.

The example dialogue shows that the use of language to describe a concept is playing at least two functions. First, the language is establishing meaning about the reality, the design work. Norms that the architectural discipline has imposed on how design concepts should be displayed linguistically are likely to be impacting the way that Petra is describing her design concept. Second, the language sets up expectations about architectural designing itself. In this dialogue, the affirmation of these expectations is done primarily by Quist. One could conjecture that there would have been a multitude of ways to describe the site and layout concept for the building. There is no one best way that she could have communicated her design concept.

This is exactly the point. The concept that Petra tries to communicate is her enactment of a set of principles for establishing the meaning of the design work available to her. In presenting her concept, she injects her voice within the voices of the architectural designers who preceded her. She has selected a certain perspective by which to communicate and enact her design concept. Her design concept sits within a frame bound by her selection of voices. Schön described the end-result of aggregation as 'types'; various types facilitate reasoning about design situations once those situations have been made into a coherent frame (Schön 1988). The linguistic pattern of aggregation captures and frames these voices into the designed work.

To gauge how language may have some regulatory function on what can be enacted, it is necessary to take into account both an individual designer's personal commitment and autonomy in speaking and writing and the situational and socio-cultural context within which the design text is produced. That is, it is worth noting that the production of the text almost always takes place in some socio-cultural context. The producer of the design text is never speaking alone, even when, for example, writing in a personal journal. In giving an account of design, the author is always referring to past experience and others' designs. The author can be thought of as borrowing from a collective voice. Whether an author wishes to address a group or reflect upon a personal thought, the author borrows a voice from a collective group. This collective group could be a school of design; it could be the positions of the stakeholders; it could be a reference to the design brief or the program; it could be a reference to the way that others design. That is, for design, the voices speak of the driving forces that shape design practice and the designed work. The voices are ideational and imaginative, constricting and delimiting. They constitute sources for the assemblage of a frame that fixes and orders design practice and the designed work.

These points offer insight into the first way in which language enacts design. A purpose of the language is to establish a coherent frame of these voices. As opposed to a merely passive purpose, namely the passive transmission and reception of information, the language serves an active role in generating new meaning. Embedded in the language is information regarding the designed work; the language is attempting to communicate, advance, and frame the designed work into a conceptual structure. However, this speaking voice is not an autonomous voice. The language is an aggregation of voices. Language use in design aggregates experiences, personal and other, where these experiences are carried through 'voices'.

Michael O'Toole describes this aggregative referencing of experiences as a form of Bakhtinian intertextuality. In making the references, "the writer/architect is saying to the viewer 'Nudge-nudge … look at my clever reference here to Stonehenge, or Palladian villas, or St Peter's in Rome, or the Pompidou Centre in Paris … It is up to you to enrich the meaning further by your knowledge of that building, its uses, its tradition, its local cultural significant, etc'." (O'Toole 2004, p. 21) Designing, like writing, involves the selection and combination of base materials to form a coherent work, a process I will call aggregation. In language, the base materials are words. In design, the base materials might be bits and bytes as with digital media, doors, walls and windows for architecture, and wood, steel and plastic for industrial design. O'Toole quoting Jørn Utzon summarizes aggregation succinctly. The designer in aggregating the base materials "must have in mind that they make a whole or an expression of some kind."

Aggregation is associated with the collection of the materiality in such a way that a coherent frame about the designed work is possible. Words are the material of aggregation; the frame is its effect. The linguistic pattern of aggregation of words frames the designed work into a conceptual structure. As Elizabeth Grosz describes aggregation in a text:

> A text is not the repository of knowledges or truths, the site for the storage of information … so much as it is a process of scattering though; scrambling terms, concepts, and practices; forging linkages, becoming a form of action. A text is not simply a tool or an instrument; seeing it as such makes it too utilitarian, too amenable to intention, too much designed for a subject. … Texts, like concepts, do things, make things, perform connections, bring about new alignments. … Ideally they produce unexpected intensities, peculiar sites of indifference, new connections with other objects, and thus generate affective and conceptual transformations that problematize, challenge, and move beyond existing intellectual and pragmatic frameworks. (Grosz 2001, pp. 57–58)

Latent Semantic Analysis

Language Theory

The computational linguistic tool we will use to reveal the aggregation of words is latent semantic analysis. Latent semantic analysis (LSA) (Landauer 1999) is a key computational linguistic tool to examine empirically the performative operator of aggregation. Latent semantic analysis models linguistic meaning as comprised of distributed relations between words in a large corpus. LSA constructs a high-dimensional space using the mathematical technique of singular value decomposition. The distributed relations are mapped in this high-dimensional space to produce a latent semantic structure from major associative patterns found in the underlying representation (Deerwester et al. 1990).

Before proceeding into the specific theoretical aspects of latent semantic analysis, we should distinguish between deep linguistic processing of natural language and statistical natural language processing, the class of computational linguistic algorithms within which latent semantic analysis belongs.

In the deep linguistic processing model, understanding natural language, the language that is being used to write this text, is based on the computational implementation of complex, linguistically designed grammars. Historically, these were based on formal ontologies and theories on how the brain organizes concepts (facts about the world). Modern theories of linguistics, such as Halliday's Systemic-Functional Linguistics (SFL) Theory, are working to provide theories and formalisms for the accurate mapping between written and spoken utterances and semantic representations in a declarative and formal way.

Deep linguistic processing has its roots in early artificial intelligence research to enable computers to use natural language, what is often called 'classical natural language processing'. Early approaches to natural language understanding were generally considered to require two competencies:

- Parsing the language into constituent components
- Having knowledge of the purpose of the constituent components, which requires:

  - Lexical knowledge
  - Syntactic knowledge (structure)
  - Semantic knowledge
  - Pragmatic knowledge (how language is commonly used in a context)
  - World-knowledge (general knowledge)

Let us suppose that we want to understand the clause *Jan gave Lynn a book*. Many approaches can be adopted, and I will only discuss two general methods to provide a flavor for the classical natural language processing approach. The ontological approach categorizes clauses as actions, states, and state changes, and tries to find the underlying causality to fit words into a structure for the communication of an idea. The clause *Jan gave Lynn a book* is an action clause, *The ball is red* is a state clause, and *Jan grew an inch* is a state change clause. The ontological approach is rather limiting given its simplistic model of language. In the decompositional approach, we look at the concepts associated with each word rather than the syntactic structure of the sentence. We attempt to elaborate the concepts and fill the words into the slots for the concepts. Since we know that the concept of *gave* requires a giver, a recipient, and an object transmitted, we would try to interpret this sentence by slotting *Jan*, *Lynn* and *book* into the appropriate slots for the concept of *gave*. One of the more well-known implementations of this approach is Marvin Minsky's frames method (Minsky 1975) which prescribes a data structure for representing stereotypical information. Each slot in the frame has restrictions. A system of frames becomes common knowledge. The challenge of applying frames was the ill-defined notion of slots; as there was no principle for defining slots, it became possible (necessary?) to codify the entire world in the slots in order to understand unusual sentences such as *Time flies like an arrow*. Should I interpret this sentence as an imperative, telling me to *time* these *flies* as I would *time* the flight time of an arrow? Or is the interpretation metaphorical about how *time* passes quickly like an *arrow* which *flies* through the air? Or is the speaker proclaiming an unusual concept combination, *time flies*, who *like*, as in adore, *arrows*? Roger Schank developed a similarly styled natural language understanding system based on the notion of conceptual dependency (Schank and Larry 1969). The conceptual dependencies derive from a network of language-free dependent concepts which also have unambiguous senses. The dependency network tries to represent a piece of natural language text in language-free terms. Again, the system is hobbled by the need for general knowledge and the difficult problem of defining language-free concepts which are unambiguous and have little if any crossover with other concepts. Schank's system also could not handle ambiguous phrases easily without resorting to a notion of the most probable intended meaning.

In contrast to deep linguistic processing and the classical natural language processing approaches, statistical natural language processing (statistical NLP) makes fairly limited use of designed grammars and *a priori* known semantic meaning. In fact, statistical NLP attempts to learn models of language from examples in a very large corpus of natural language data using the formalism of probability theory. Statistical NLP algorithms attempt to estimate the parameters of these probabilistic models of language so that that they can be used in the processing of natural language using supervised and unsupervised machine learning approaches. Generally, they depend on the availability of linguistic databases that include rich linguistic annotation to corpora of natural language text such as the Penn Treebank distributed by the Linguistic Data Consortium. Thus, on the one hand, there are grammatical, rule-driven, symbolic approaches to understanding natural language; these are the deep linguistic processing and classical NLP approaches. On the other hand are the statistical NLP methods. Modern computational linguistics research combines both approaches in order to achieve high throughput language understanding and generation.

Latent semantic analysis takes the even more challenging view that it is possible to induce knowledge from local co-occurrence data of words from a large corpus of representative text. LSA is unique in its method to analyze text; in its analysis, there is no consideration of word order or syntax. Different from many other statistical NLP approaches, there is no prior linguistic model or any probabilistic model that it tries to fit parameters to. The principal advantage of LSA over other computational linguistics techniques is LSA's examination of context instead of individual word meanings. The baseline theory for LSA is that word choice and word and document meaning patterns emerge from the entire range of words chosen and words that were not chosen in a wide variety of texts. LSA is capable of inferring deeper relations of semantic knowledge, hence the term latent semantic.

This assertion has some additional support from cognitive science (Gagné 2000). Cognitive psychological research on how noun–noun concept combinations are interpreted has established that humans can interpret concept relations through property-mapping and relation-linking if semantic relation mapping 'fails'. In property-mapping, we map the property of one concept onto the other. In relation-linking, we believe that the relation of a concept as 'being for' the other concept.

Let us suppose that we did not know about the concept *book worm* as a person who avidly reads books. In a property-mapping linking, we might think that *book worm* refers to a *worm* that has the shape of a *book*. In the relation-linking, we might think that a *book worm* is a *book* for *worms* or a *book* that has *worms*. These possible interpretations are not necessarily generated by any *a priori* semantic relation between *book* and *worm*. Rather, they arise due to relations that were generated through the occurrence of the words *book* and *worm* with other words that we have encountered. These weak, latent relations between words are the sort of relations that latent semantic analysis tries to impute.

That LSA examines context removes the obfuscation created by noise[2] in oral and written text, and scales up to deal with very large corpora. Many computational linguistics tools which depend on explicit semantic meaning or expected grammatical forms degrade in performance when the data set is noisy or semantic meaning is ambiguous and polysemous. As such, LSA is unique in its method of analyzing text in that there is no consideration of word order, syntax, or grammar[3]. However, that LSA can function effectively independent of grammar and per word (or word group) semantic meaning is of particular import when we need to analyze informal design text such as transcriptions of design conversations, e-mails, and blogs – that is, the types of text that are produced in parallel with design activities. Oral and informal written communications in design tend to be especially noisy. It is through these types of informal design documents, rather than say manuals of designed works, that we anticipate to locate the performative realization of the designed work and the design process by language.

While originally developed to identify contextual meanings of documents as a method for full-text information retrieval (Deerwester et al. 1990), Landauer theorizes, somewhat controversially, that LSA provides a computational model of language acquisition and acquired knowledge (1997). Based on a study showing the capability of LSA to extract and induce knowledge from text at rates similar to or better than humans, Landauer and Dumais concluded:

> It is supposed that the co-occurrence of events, words in particular, in local contexts is generated by and reflects their similarity in some high-dimensional source space. ... We hypothesized that the similarity of topical or referential meaning ("aboutness") of words is a domain of knowledge in which there are very many indirect relations among a very large number of elements and, therefore, one in which such an induction method might play an important role. (Landauer and Dumais 1997, p. 234)

Second, they claimed that the higher-order associations contained in the usage of words may generate metalinguistic social phenomena, such as "self", "personal identity" and "social group perception". Thus, LSA is an important computational tool to explore theoretically and empirically align the existence and choice of

---

[2] Noise in text can result from several factors including frequent ungrammatical sentence structures such as incomplete sentences, unusual word choices, misspellings, or highly specialized jargon.
[3] This point has been a significant critique of LSA, particularly in its application to grading student essays (Hearst 2000).

Fig. 3.1 (*top*) The singular value decomposition of the matrix **X** into left and right singular matrices **U** and **V** and the singular matrix **S**; (*bottom*) the approximation of **X** as $\hat{\mathbf{X}}$ by retaining the first k largest singular vectors

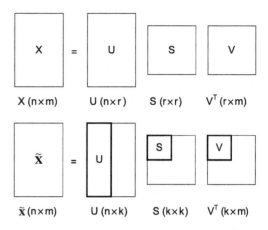

words to metalinguistic phenomena. In particular, since design sometimes involves the production of concept relations which do not yet exist, that LSA could deduce relations of semantic meaning without the need for known thesaural associations is an advantage in inducing constituent concept combinations that do not occur frequently in everyday discourse. For our purposes, LSA will serve as a computational tool to examine how word choice patterns in design text aggregate indirect and perhaps opaque relations between the complex relations among words from which design concepts can be framed.

The mathematical foundation for LSA lies in singular value decomposition (SVD), a matrix approximation method for reducing the dimensions of a matrix to the most significant vectors[4]. The singular value decomposition of the matrix **X** of dimension $n \times m$, $m < n$, is defined as $\mathbf{X} = \mathbf{USV}^T$ where **U** $(n \times r)$ and **V** $(r \times m)$ are the left and right singular matrices (orthonormal), respectively, and **S** $(r \times r)$ where r is the rank of **X** $(r \le m)$ is the diagonal matrix of singular values, as shown in Fig. 3.1 (top). The abstract matrices **U**, **S** and **V** contain all of the relations necessary to reproduce the original matrix **X**. In latent semantic analysis, the words are normally represented in the rows of **X** and the documents in the columns.

SVD yields a simple strategy to obtain an optimal approximation for **X** using smaller matrices. If the singular values in **S** are ordered descending by size, the first k largest may be kept and the remaining smaller ones set to zero. The product of the resulting k-reduced matrices is a matrix $\hat{\mathbf{X}}$ which is approximately equal to **X** in the least squares sense and of the same rank. That is, $\hat{\mathbf{X}} \cong \mathbf{X} = \mathbf{USV}^T$. This matrix is called the k-reduced matrix, shown in Fig. 3.1 (bottom). The number of singular dimensions k to retain is an open issue in the latent semantic analysis research[5]. Based on empirically derived results, the common practice is to retain dimensions 2 to 101.

---

[4] See (Strang 1988) for a derivation of singular value decomposition.

[5] Probabilistic latent semantic analysis (Hofmann 2001) corrects this problem with LSA using a maximum likelihood estimation of the latent variables. There is a slight improvement in document retrieval accuracy at the cost of increased computing time to compute the maximum likelihood estimation.

A special property of SVD is its ability to re-represent the original matrix $\mathbf{X}$ as a vector of uncorrelated factor values. This is achieved by the singular matrix $\mathbf{S}$; $\mathbf{S}$ 'scales' and 'stretches' the matrices $\mathbf{U}$ and $\mathbf{V}$ such that the word and document vectors, respectively, from $\mathbf{X}$ are re-represented in the k-reduced space. The scaling of the document vectors of $\mathbf{X}$ (that is, the columns of $\mathbf{X}$) is shown in the mathematical proof that the rows of the matrix $\mathbf{VS}$ are coordinates of documents in the k-reduced space. The similarity between any two documents is the dot product between two column vectors of $\tilde{\mathbf{X}}$. It turns out that the matrix $\tilde{\mathbf{X}}^T \tilde{\mathbf{X}}$ is the dot product of the document vectors $\mathbf{VS}$. Since $\tilde{\mathbf{X}} = \mathbf{USV}^T$, then $\tilde{\mathbf{X}}^T \tilde{\mathbf{X}} = \left(\mathbf{USV}^T\right)^T \mathbf{USV}^T = \mathbf{VS}^T \mathbf{U}^T \mathbf{USV}^T = \mathbf{VS}^T \left(\mathbf{U}^T \mathbf{U}\right) \mathbf{SV}^T = \mathbf{VS}^2 \mathbf{V}^T$ since $\left(\mathbf{U}^T \mathbf{U}\right) = \mathbf{I}$.

That is, we can interpret the rows of the matrix $\mathbf{VS}$ as the coordinates of 'scaled' documents in the k-reduced latent semantic space. This useful property of SVD means that it is possible for documents with slightly different patterns of occurrences of words to be mapped into the same vector of factor values. The interpretation of this (mathematical) property is that the matrices $\mathbf{U}$, $\mathbf{S}$ and $\mathbf{V}$ capture the reliable and relevant semantic patterns to define the concept space of the documents. This concept space is an abstract linguistic framing for the design concept, and the approximated matrix $\tilde{\mathbf{X}}$ is presumed to represent the most vital patterns of semantic relations in the original matrix $\mathbf{X}$. Column vectors in $\tilde{\mathbf{X}}$ represent the latent semantics of a document imputed through SVD instead of the explicitly inscribed semantics. These latent semantics allow us to overcome the problem of insufficient context from which to establish the semantic relations in text. The column vectors of $\tilde{\mathbf{X}}$ are the basis for calculating aggregation.

Since LSA does not depend upon semantic meaning but instead induces semantic relations from terminological conventions in a representative text, the representation of text is based on a word-by-document matrix. In the word-by-document matrix, the words are represented in the rows of the matrix $\mathbf{X}$ and the documents in the column. The value of each cell is a measurement of the amount of 'content' of the word in the respective document. The measurement of content is known as term weighting.

The word-by-document matrix is a standard format of text representation in the field of full-text information retrieval. One of the major achievements of statistical natural language processing, and perhaps the most practical outcome of the research in natural language processing in general, is in the area of full-text information retrieval. This is the sort of retrieval everyone does routinely when using a search engine such as Google. Full-text information retrieval engines index documents using words or phrases found in the documents. The retrieval algorithm tries to find the closest matching document to the query entered by the user. In order to calculate the score of the closest matching document to the query, the system maintains statistical information about the frequency of occurrence of the word in individual documents and about the frequency of occurrence of the word across all documents in the corpus. Based on these frequency measurements, the system then calculates the match between the query and documents where the query terms appear.

One of the reasons why full-text information retrieval systems have been commercially successful is in the simplicity of the mechanism by which natural language is processed and stored as a *vector space model*. By combining the vector space model for each document in the corpus, one obtains the word-by-document matrix. In the vector space model, individual documents and queries are represented as vectors in a word space where the vector weights contain measurements known as a term weighting scheme. The term weighting scheme attempts to measure, statistically, how much 'content' is contained in the respective word. The term weight is usually a value between 0 and 1. A weighting scheme is composed of three different types of term weighting: local, global, and normalization. The term weight is given by:

$$L_{ij} G_i N_j \qquad (3.1)$$

where L_{ij} is the local weight for term i in document j, G_i is the global weight for term i, and N_j is the normalization factor for document j. The local and global weights attempt to balance the notion that words which appear frequently across all documents tend not to be discriminating but they need to appear in more than a few documents and if they appear frequently in any one document, then that word must be indicative of the document's content. The normalization factor compensates for discrepancies in the lengths of the documents. If no weighting is applied, the value is 0 means that the word does not appear in the context and the value 1 means that it does.

To balance the effect of word frequencies in the text, log-entropy term weighting may be applied to the original word-by-document matrix $\mathbf{X}$. Let us call this log-entropy word-by-document matrix $\mathbf{F}$. Each cell value of $\mathbf{F}$ is then scaled according to the following equation:

$$\mathbf{F}_{ij} = L_{ij} G_i \qquad (3.2)$$

where

$$L_{ij} = \log_2\left(\mathit{tf}_{ij} + 1\right) \qquad (3.3)$$

$$G_i = 1 - E_i \qquad (3.4)$$

$$E_i = \sum_{j=1}^{m} \frac{p_{ij} \log_2 p_{ij}}{\log_2 m} \qquad (3.5)$$

$$p_{ij} = \frac{\mathit{tf}_{ij}}{\sum_{j=1}^{m} \mathit{tf}_{ij}} \qquad (3.6)$$

and tf_{ij} is the frequency of occurrence of term i of n words in document j of m documents. This term weighting scheme gives very frequent words low weight and assigns large weight for infrequent words.

Table 3.2 Word by document matrix for the sample text

	d_1	d_2	d_3	d_4
w_1 shop	1	0	0	1
w_2 would	1	0	0	0
w_3 prefer	1	0	0	0
w_4 trolleys	1	0	1	0
w_5 carry	1	1	0	0
w_6 merchandise	1	0	0	0
w_7 parents	0	1	0	0
w_8 want	0	1	0	0
w_9 young	0	1	0	0
w_{10} children	0	1	0	0
w_{11} shopping	0	1	0	0
w_{12} seat	0	0	1	1
w_{13} make	0	0	1	0
w_{14} belt	0	0	1	0
w_{15} safe	0	0	1	0
w_{16} comfortable	0	0	1	0
w_{17} advertise	0	0	0	1
w_{18} brand	0	0	0	1

As an example of latent semantic language model in practice, here are four sentences from four fictitious design documents. Suppose that d_1 is from a brainstorming session, d_2 is from a customer and user needs survey/interview, d_3 is from the description of a prototype, and d_4 is from a marketing brochure. Each unique content-bearing term is italicized.

d_1: *Shop would prefer trolleys* to *carry merchandise*
d_2: *Parents want* to *carry young children* while *shopping*
d_3: *Make seat belt safe* and *comfortable* on *trolleys*
d_4: *Advertise shop brand* on the *seat*

The word by document matrix for the sample text is shown in Table 3.2.

There are a couple of important items to note about the word by document matrix. First, the matrix is quite sparse; that is, the documents in general do not share many words in common. The word by document matrix for an arbitrary set of documents will be extremely sparse. The most significant characteristic to note from the example is the relatively 'weak' explicitly inscribed semantic similarity between the sentences. It is through the aggregate co-occurrence of the words shop, carry, and trolleys in the text that the latent semantic connections between these words appear. They become semantically similar not through *a priori* prescribed semantic similarity but through terminological convention.

Second, given the sparsity, most documents would not normally appear to be related to one another using the standard cosine similarity measurement of the

Fig. 3.2(a) The singular value decomposition of matrix **X** into abstract matrices **U**, **S** and **V** for the sample text

$$X = USV^{\mathrm{T}}$$

$$
U = \begin{bmatrix}
-0.3570 & -0.0347 & -0.3689 & 0.3283 \\
-0.2425 & 0.0485 & -0.3053 & -0.3053 \\
-0.2425 & 0.0485 & -0.3053 & -0.3053 \\
-0.4380 & -0.1792 & -0.0391 & -0.2937 \\
-0.3753 & 0.3669 & -0.0850 & -0.1026 \\
-0.2425 & 0.0485 & -0.3053 & -0.3053 \\
-0.1328 & 0.3184 & 0.2204 & 0.0644 \\
-0.1328 & 0.3184 & 0.2204 & 0.0644 \\
-0.1328 & 0.3184 & 0.2204 & 0.0644 \\
-0.1328 & 0.3184 & 0.2204 & 0.0644 \\
-0.1328 & 0.3184 & 0.2204 & 0.0644 \\
-0.3100 & -0.3110 & 0.2027 & 0.3688 \\
-0.1955 & -0.2277 & 0.2663 & -0.1266 \\
-0.1955 & -0.2277 & 0.2663 & -0.1266 \\
-0.1955 & -0.2277 & 0.2663 & -0.1266 \\
-0.1955 & -0.2277 & 0.2663 & -0.1266 \\
-0.1145 & -0.0833 & -0.0636 & 0.4954 \\
-0.1145 & -0.0833 & -0.0636 & 0.4954
\end{bmatrix}
$$

$$
S = \begin{bmatrix}
2.7975 & 0 & 0 & 0 \\
0 & 2.4804 & 0 & 0 \\
0 & 0 & 2.1481 & 0 \\
0 & 0 & 0 & 1.8458
\end{bmatrix}
$$

$$
V = \begin{bmatrix}
-0.6784 & 0.1204 & -0.6559 & -0.3084 \\
-0.3715 & 0.7898 & 0.4734 & 0.1189 \\
-0.5469 & -0.5649 & 0.5719 & -0.2337 \\
-0.3203 & -0.2065 & -0.1366 & 0.9144
\end{bmatrix}
$$

dot-product between the document vectors. The cosine similarity of two documents d_p and d_q is defined by Eq. 3.7.

$$\cos(d_p, d_q) = \frac{d_p \cdot d_q}{\|d_p\| \|d_q\|}. \tag{3.7}$$

For example, the cosine similarity of d_1 and d_2 is (1,1,1,1,1,1,0,0,0,0,0,0,0,0,0,0,0) · (0,0,0,0,1,0,1,1,1,1,0,0,0,0,0,0,0) = 0.1667; the maximum possible cosine similarity is 1. These two documents should be quite similar since they both discuss using *trolleys* to *carry* items including *merchandise* and *children*. The strength of LSA lies in relating documents even when words in the documents may not match exactly. The singular value decomposition of the matrix shown in Table 3.2 into the projected matrices is given in Fig. 3.2(a), and the approximated

Fig. 3.2(b) The singular value decomposition of matrix $\mathbf{X}$ into the approximation $\tilde{\mathbf{X}}$

$$\tilde{\mathbf{X}} = \begin{bmatrix}
0.6672 & 0.3030 & 0.5949 & 0.3376 \\
0.4747 & 0.3471 & 0.3030 & 0.1924 \\
0.4747 & 0.3471 & 0.3030 & 0.1924 \\
0.7778 & 0.1024 & 0.9213 & 0.4842 \\
0.8219 & 1.1089 & 0.0601 & 0.1483 \\
0.4747 & 0.3471 & 0.3030 & 0.1924 \\
0.3471 & 0.7618 & 0.7618 & -0.0441 \\
0.3471 & 0.7618 & 0.7618 & -0.0441 \\
0.3471 & 0.7618 & 0.7618 & -0.0441 \\
0.3471 & 0.7618 & 0.7618 & -0.0441 \\
0.3471 & 0.7618 & 0.7618 & -0.0441 \\
0.4955 & -0.2870 & 0.9101 & 0.4371 \\
0.3030 & -0.2429 & 0.6183 & 0.2918 \\
0.3030 & -0.2429 & 0.6183 & 0.2918 \\
0.3030 & -0.2429 & 0.6183 & 0.2918 \\
0.3030 & -0.2429 & 0.6183 & 0.2918 \\
0.1924 & -0.0441 & 0.2918 & 0.1452 \\
0.1924 & -0.0441 & 0.2918 & 0.1452
\end{bmatrix}$$

matrix $\tilde{\mathbf{X}}$ is given in Fig. 3.2(b). The highlighted portions (within the dashed-line box) of the matrices $\mathbf{U}$, $\mathbf{S}$ and $\mathbf{V}$ are the values on the first and second dimensions of the component matrices. By retaining only the first two singular dimensions, the k-reduced matrix approximation $\tilde{\mathbf{X}}$ of the original matrix $\mathbf{X}$ is calculated.

In the k-reduced form, the similarity between documents d_1 and d_2 is now 0.6003, indicating a stronger relation between a potential product concept and one customer's needs than originally computed. Also, note that although the words trolleys (w_4) and seat (w_{12}) only co-occur once in d_3, the dot product between rows 4 and 12 increased from 0.5 to 0.9290, indicating a stronger relationship between these two words. This happens because the word *shop* (w_1) co-occurs with *trolleys* (w_4) and *seat* (w_{12}), which LSA considers as a latent imputed semantic relation between them. The imputed relation would not have appeared with a simple keyword match using the original sparse word by document representation. LSA reveals semantic relations where simple keyword matching would not, essentially eliminating the latter as an appropriate tool for this type of text analysis.

Now that we have a representation of the text that allows us to impute semantic relations between words and documents, we need to measure the strength of aggregation of the framing of the design concept. Picture in your mind the text data in a very high dimension space. If you find it useful, imagine that the space is a constellation of stars, where each document is represented by a star in a constellation (of documents). The documents might orbit about a sun at the center of the constellation. In LSA terms, this abstract 'average document' is called the 'concept centroid'. The concept centroid is an abstract semantic representation of the design concept. Mathematically, the centroid is computed by finding the average value of each row in the k-reduced matrix. The centroid is of dimension $n \times 1$. Define c as the centroid of the document corpus.

Now, imagine we need to calculate the density. There are two ways we could go about this. One way is to calculate how far the documents are away from the center. The Variation in Semantic Choice (Eq. 3.8) metric indicates the closeness of each text to the average document. This metric calculates the magnitude of deviations of the s-dimensional radius (as a Euclidean distance) of each document vector in the LSA space. Conceptually, the centroid of the document vectors represents the aggregation into a coherent concept; dilution of that aggregation would manifest through the variance of the radii and distance to outlying texts.

$$\sigma = \frac{\sqrt{\sum_{j=1}^{m}(s_j - c)^2}}{m} \qquad (3.8)$$

where s_j is the document vector **VS** representing the j-th document and m is the total number of document in the corpus.

Another way to measure the density is to measure the average coherence between documents. The Semantic Coherence (Eq. 3.9) metric measures the (mutual) aggregated semantic similarity of the texts and is a measure of the quality of coherence (Dhillon and Modha 2001). The semantic coherence is the similarity with which all texts represent the various voices referred to in forming the concept and in allowing the reader to form an interpretation of the concept. Analytically, the norm calculates whether all singular vectors for a corpus are identical. It is as if all the voices were expressing the same subject although possibly with slightly different grammatical styles.

$$\chi = \left\| \sum_m s_j \right\| \qquad (3.9)$$

Computational Implementation

Our aim is to examine how words frame the concept in a higher-order linguistic space. What we need to imagine is the existence of a hyper-space of words in which the position of the words in the space is located by its vector representation (vector-space model). The intuition is that LSA will allow us to see the innumerable word constituents of concepts aggregate, eventually producing a system of design concepts. The latent semantic approach to analyze for aggregation proceeds as follows:

1. Use a natural language parser (such as JavaNLP[6]) to extract words and phrases from design texts.
2. Create a word by document matrix which counts the frequency of occurrence of each word in each document from Step 2.
3. (Optional) Use log-entropy scoring on the matrix from Step 2 to filter the word list to a set of discriminating terms. A useful cut-off is to include terms with an average log-entropy greater than or equal to the mean log-entropy across all words.[7]
4. Apply LSA (singular value decomposition) on either the raw frequency or log-entropy matrix. SVD is available in most mathematical computing software.[8] For extremely large matrices (i.e., on the order of a million rows or columns), try SVDPACK (Berry 1992).
5. Compute the k-reduced approximation of the original word by document matrix. It has been experimentally determined in various research projects that retaining up to the first 300 most significant singular dimensions provides the best approximation of the latent semantics. For these analyses, dimensions 2 to 101 were retained.
6. Calculate the metrics of aggregation based on the k-reduced matrix.

Generating the word by document matrix is generally a simple matter of text processing. In preparing the text (Step 1) to generate the matrix (Step 2), the natural language text is tokenized into individual words. Given the availability of fairly efficient and accurate part-of-speech tagging systems such as the Stanford Java Natural Language Parser, extraction of noun and verb phrases is also possible. However, it is quite possible to quickly tokenize a large corpus of text with just a few lines of code in PERL and the split function: `my @words = split(/\s/,$_);`

Once the words from the documents have been tokenized, one typically throws away stop words or non-content-bearing words. Various stop word lists are available on the Web. Then, one writes some software code to create a word-by-document matrix which counts the frequency of occurrence of each word in each document in the corpus.

Again, this can be done quite simply with PERL. Suppose the text for each document in the corpus was stored in a database where each tuple contained a single document and the text in the document were stored as a text field. We could write a PERL function to count the number of times a word appeared in the text field using the following PERL statements:

---

[6] You can download JavaNLP from the Stanford JavaNLP Project Home Page at `http://nlp.stanford.edu/javanlp/`.

[7] In practice, I have not found using log-entropy weighting to be useful to "throw away" words. The so-called "junk" words seem to serve the purpose of relating text together when their relations are not definable through known semantic relations.

[8] The singular value decomposition function in MATLAB® is `svd`. Use the form `[U,S,V] = svd(X,0)` to compute the "economy size" decomposition.

```
while (my @row_ary = $sth->fetchrow_array()) {
    # Count the frequency of word in the text field
    foreach my $word (@wordlist) {
        my $match = 0;
        chomp $word;
        while ($row_ary[0] =~ m/.*\b($word)\b.*/gi) {
        $match++;
        }
    }
}
```

Sample Study

Let us now examine aggregation in design text. To do this, we link the performative operator of aggregation acting upon design text to the production of shared understanding in a group design situation. Here, empirical studies (Dong 2005; Dong et al. 2004; Hill et al. 2002) have relied on the study of design teams since language-based communication is integral when working across mental landscapes. In the empirical investigations of group design, we were interested in addressing the question how it is that design teams come to 'know' what is it that they're designing. The hypothesis is that the similarity of language use bridges indirect relations among components of knowledge stored in each designer's mind, leading to a constructed shared mental representation of the designed artifact. This hypothesis relates a cognitive enactment of aggregation which is that shared understanding is a cognitive effect of aggregation. Since the documents were authored by various team members, each contributed individual knowledge to the artifact's function, behavior, structure, and meaning. Words and phrases used by designers during the design process contributed to a narrative which captures, among many items, personal experience, functional specifications, negotiations, and resolutions. Design specifications and solutions in text relate more than numbers and formulae; they reflect the conflicting interests and resulting reconciliation and shared agreements of the design team members. The semantic coherence of their language-based communication is a metric of the congruency of not only their joint thinking but also their linguistic framing of the complex interdependencies between often competing interests into a coherent design concept. The events and elements of the design process are narrated in a manner that inter-relates them throughout the design process, signaling a continuously developing synthesis.

The studies calculate two metrics, the Variation in Semantic Choice (Eq. 3.8) and Semantic Coherence (Eq. 3.9), to assess congruent communication. It is important to note that we utilized latent semantic analysis to measure textual coherence – not textual cohesiveness. Cohesiveness in text (lexical cohesion) is based on the pragmatics of grammar, or "the ties that bind a text together" (Halliday and Hasan 1976) whereas textual coherence relates to the thematic consistency. That is, a coherent text presents (a) unified concept(s) "about the speaker's or author's purpose" (Fillmore 1974).

As an example of the contrast between textual coherence and cohesiveness, consider the two excerpts from design documents. The first (A) comes from the original project proposal by a team member, whereas the second (B) is the product description from the second draft of the product mission statement.

A: There are very many different types of medical products that are bought by consumers and over their lifetime consumers will purchase at least one prosthetic device.
B: To provide a comfortable and stylish assistive walking device for active and sporty adults.

Even though there is weak evidence of textual cohesiveness between the two versions, there appears to exist a consistency of theme, medical devices. In fact, if all the documents authored by a team were collapsed into one, the grammar and structure of the various authors would likely not be textually cohesive. LSA is an apparatus to explain aggregation through a direct measurement of textual coherence.

Studies of design documentation were based on material from student product design teams. The documents were collected over a 15 week period and consisted of e-mail, mission statements, customer feedback, product testing and engineering memos, presentation materials, and personal and group reflections on the team's execution of the process and product at each stage of the product development process. The semantic space was constructed by collecting on the order of 10^2 documents per team for which each document contained on the order of 10^1 to 10^3 words.

The projects were typically consumer products with low complexity. While the students self-selected the projects, the academic staff guided the students in choosing products to enable learning the product development process and engineering design rather than focusing on specific technological challenges. Large amounts of customer feedback and user input are encouraged through the process and generally teams do a quality job of this. By studying student design teams instead of industrial firms, we were able to remove some external factors that would greatly affect the team's performance, such as the market conditions and variations in the product development process (i.e., differences in process that may lead to varying outcomes).

What we found is that the highest performing teams, as ranked by expert (professional product designers and academic staff) assessments of the final product and team process, produced the most coherent design documents. Table 3.3 reports on the relationship between the expert ranking of the team's design performance and a scaled value for each team's variation of semantic coherence in a study we conducted on eight student design teams (Dong et al. 2004). The scaling is calculated as ten times the fraction of the respective team's variation in semantic coherence to the highest variation for the data set. For example, if the worst performing team had a variation in semantic coherence on the order of magnitude 10^1, then the scaled variation in semantic variance is $10 \times (10^1 \div 10^1) = 10$. On the other hand, if a team has a variation in semantic coherence on the order of magnitude 10^0, then the scaled

Table 3.3 Relationship of team rank to scaled variation in semantic coherence

Team	Rank	Scaled Variation in Semantic Coherence
A	1	1
B	2	3
C	3	2
D	4	1
E	5	3
F	6	2
G	7	7
H	8	10

variation in semantic variance is $10 \times (10^0 \div 10^1) = 1$. We note in particular that the poorest performing teams, Teams G and H, had a scaled variation in semantic coherence more than twice the others. There is a distinguishable difference in semantic coherence between well-performing teams and poorly-performing teams.

Examining the comments that judges made relative to high performing teams and low performing teams can have further insight into the analysis. The following are comments given to a high performing team:

1. "Direct, Specific and Clear"
2. "Well thought out process to determine customer needs and generate concepts."
3. "Good rigorous methods to organize all the features and combinations."

All told, the comments about the high performing teams in general, indicated a positive impression of the team's design concept and how the concept was enacted. When contrasted with comments about low performing teams, a clear difference can be seen in how this team's designed concept was created and enacted into the designed work. Comments about a low performing team are as follow:

1. "Too broad. It would be difficult to screen potential product designs. Seems more like a mission statement for a broad range of products that provide similar services."
2. "Focused too much on usability issues and not enough on functional requirements."
3. "OK. But did not explore problem."

The sentiments reflected in the judges' comments relate to problems in the teams' design process including the lack of a clear concept for the designed work, distracting focus on non-key issues, and a design approach that focused on a single solution too early in the process, which is fairly typical of inexperienced designers. These sentiments, we think, accurately describe a situation for which the language of design failed to aggregate. In the terminology of Austin, the "Infelicity" of the performative operator of aggregation in not achieving a coherent design concept does not mean that aggregation is not present in the 'unhappy' design text.

Rather, it means that there will exist variations in how or even whether the performative operator achieves its effects. The insight is that this operator can provide an indicator for the conditions of use of language in design which parallel successful and non-successful designing.

Low-performing teams struggled to clearly state and define their product's objectives. Consider the mission statements from two teams, one high-performing team (Team D) and one low-performing (Team G). The number in parenthesis indicates the version of the team's mission statement.

Team G:

(1) Eliminate all adverse effects or deaths resulting from mis-filled prescriptions.
(2) Product-based solution to eliminate all adverse effects or deaths resulting from mis-filled prescriptions.
(3) An assistive device that improves the ergonomic design of the crutch for users with lower limb limitations.

Team D:

(1) A system that matches people resource to projects in an optimal manner so as to optimize efficiency and effectiveness of the resource management process for these projects.
(2) A system that matches resources to projects in an optimal manner so as to optimize efficiency and effectiveness of the resource management process for these projects.
(3) A system that will recommend the "right" people resources for projects according to the needs of project, as well as those of the resources themselves. This facilitates a more efficient and effective resource allocation process.
(4) A specialized system that augments the process in which project managers find resources for their projects and resources find projects for themselves.

Whereas Team D presents a mission statement that iteratively broadens, narrows and precisely defines the mission of the product, Team G struggles to define a product, initially planning to develop a product centered around pharmaceutical prescriptions and eventually settling on assistive walking devices.

Lack of aggregation toward a coherent concept also created confusion for Team H. In reflecting on product ideas that each team member presented early on in the process, one member wrote, "… everyone has a different understanding what the paper says." Consequently, "Coming up with a product idea was really difficult." We believe this lack of coherent product statement and direction is inscribed in the design text.

There is an alignment between the failure for the concept to cohere and the design process. One team member in Team E (low-performing) wrote, "Trying to channel the focus of the project has long been the biggest headache. My 'engineering' opinion of choosing a specific direction then widen the concept ideas around the direction is right in conflict with the rest groups' 'business' opinion of 'letting the customer need' let [sic] us what direction to go." Shared norms for decision-making were a stumbling block for Team F. One team member wrote, "I think our team took too much time making decisions – we should have been quicker to implement decisions. [Some team members] provided valuable insights and direction but dilly dallied a lot of time making decisions." Regarding a particularly difficult decision the team made with respect to switching product directions, one team member was apparently never fully convinced. "I strongly disagreed initially but had to go with the team. However I think the aim of the [project] is the 'process' of product development and less focus is on the 'actual product' so I thought we should have continued with the original concept."

The actual numeric values of semantic coherence will, of course, vary depending upon the teams and the type of design project (e.g., routine, innovative, breakthrough). As such, they are not reported and are only meaningful when compared across similar design situations. Comparisons of design teams within companies, and possibly companies of similar size in the same industry, are reasonable, but comparisons between teams in differently sized companies or under differing sets of pressures may lead to unstable results.

While the above studies examined design documents, we would also like to analyze what happens when designers talk about designing the concept in real-time. Studying synchronous language-based communication turns out to be more problematic. A meaningful computation of variation in semantic choice and semantic coherence is not possible. For verbal communication, each 'document' is generally a set of short utterances, most of which consist of too few words to establish large enough contexts for LSA. Another approach is required.

Using an intuition about conversations, we could expect that utterances close to one another should be thematically similar. That is, if I say, "The visualization should exhibit a strong contrast in colors", your reply could be, "To make it easier to distinguish objects in the scene" but probably not "I don't like pears", although such diversions are not unheard of during experiments of group design nor in reality. Although these two utterances do not share any explicit semantic similarity, they do share an imputed semantic coherence since it is altogether possible that the words visualization, contrast, colors, objects, and scene would appear together when discussing the visual design of a data visualization to make it easier for viewers to discern objects. Translated computationally into LSA, if we calculate the aggregate average of the semantic coherence between utterances (Eq. 3.7) which are one utterance away, two utterances away, and so forth, then we should be able to plot a pattern of semantic coherence that, on average, shows coherence between distally close utterances. The presumption is that if the language use were coherent, then there should exist a mostly orderly mapping between semantic coherence and distance between utterance boundaries. An utterance boundary is defined by the turn-taking between speakers, that is, when one speaker stops talking and the next speaker begins. The intuition is that utterances close together should be similar, utterances further apart less so. This intuition is a similar to a method developed to examine thematic coherence in written text (Foltz et al. 1998). Foltz and colleagues found in their study that there should exist a mostly ordered relation between semantic similarity and distance between paragraph and chapter boundaries in books.

This ordered relation can be revealed by examining the coherence between any two utterances (communicative acts) as a function of the distance between the utterance boundaries. In summary, the tactic is to calculate the average coherence between utterances which are one utterance away, the average coherence between utterances which are two utterance boundaries away, and so forth, to expose the structuring of language over the entire conversation. I would argue that the patterns of thematic coherence display the aggregation of lexicalized concepts to form the concept. In group design, the alignment of ideas leads to the construction of a design concept. Depending upon the pattern of increase and decrease in coherence as a function of utterance boundary distance, how utterances imbricate semantically may account for the effect of aggregation in verbal design communication.

In order to validate our computational approach, it is necessary to compare the results with those generated through hand-done analyses. While we are not looking to automate the human analyses, it is important that our computational results are consistent with prior analyses. To examine the formation of coherence in design conversations, we chose standard transcripts from two widely studied projects in design science research. The following two data sets are from controlled, laboratory studies of group design in which groups of designers working together are tasked with the conceptual design of an artifact that satisfies a set of requirements. That is, their design conversations specifically deal with the production of a 'new' design concept. For these data sets, the distance between utterance boundaries is calculated as the number of utterances between turns by different speakers. Each utterance may consist of multiple complete sentences or incomplete thoughts.

The first was a transcript from the mountain bike backpack design problem from the 1994 Delft Protocols Workshop. Recorded on videotape and a written transcription of their conversation over a two-hour period, three professional designers, identified as Ivan, John and Kerry, designed a fastening device that should allow a given backpack to be fastened to a bicycle (a common problem for the bicycle-friendly Netherlands!) The transcript contains 2190 raw utterances among three design students over a 118 minute period. Excluding utterances containing only noise terms such as "mmmm" and "oh", there were 1662 utterances. Participants Ivan, John and Kerry spoke 34%, 39% and 27%, respectively, of the group conversation.

The second set comes from a study by Petra Badke-Schaub and Joachim Stempfle at Universität Bamberg (Stempfle and Badke-Schaub 2002) of design thinking in teams based on their communication. There are three teams in the Bamberg Study, denoted by 1102, 2202 and 2302. Bamberg Team 1102 consisted of 6 participants (A–F) who contributed 15%, 21%, 9%, 20%, 18% and 16% of the content-bearing utterances; Bamberg Team 2202 consisted of 4 participants (A–D) who contributed 32%, 16%, 30% and 21%; and Bamberg Team 2302 consisted of 4 participants (A–D) who spoke 40%, 13%, 19%, and 28%. For brevity, we will use the notation 1102/D, for example, to refer to Participant D in Bamberg Team 1102.

Before proceeding to the results of our study, let us review the findings of other researchers who have investigated these design teams. Valkenburg (1998) took up the question of the formation of shared understanding in the Delft backpack design team. Valkenburg defines shared understanding in design as "a mutual knowledge of all team members on what they are doing, why, and how they are doing it."

(1998, p. 120) She found that the team reached shared understanding on some points of the design, but not others. The team came to a shared understanding on designing the fastening device around a best-selling external frame backpack, but not on the adjustability of the rack. Overall, however, the team came to a shared understanding of the principal conceptual aspects of the fastening device. In another widely cited study of this team, Gabriela Goldschmidt (1996) characterized John as the ideas person and as the most active in driving the direction of the team. Ivan is the process manager and the time keeper who summarizes but weakly influences the team. Kerry has the most domain knowledge and appears to make specific contributions to the functional specifications.

Stempfle and Badke-Schaub (2002) were looking for a generic model of design to explain the thinking processes of designers working in a group. By coding up the conversation of teams of designers according to their generic model, they developed a two-process theory of thinking in design teams. In first process model, the design team proposes a solution and then immediately evaluates the solution without much, if any, analysis of the proposed solution. In process two, the team generates sets of solutions, analyzes all of them, and then evaluates them for possible selection. While schooled in the formal German Verein Deutscher Ingenieure (VDI) VDI 2221 design process, only one of the teams (Bamberg Team 2202) followed the 'prescribed' process. This team analyzed a proposed solution 55% of the time after generating a solution and before evaluating for acceptability whereas the other two groups seemed to frequently propose a solution and then immediately evaluated the solution, bypassing analysis. The Bamberg researchers reported that "proceeding in the design task can be labeled from 'chaotic' (group 3) to 'planned' (group 1)" where group 3 is Bamberg Team 2302 and group 1 is Bamberg Team 1102 (Stempfle and Badke-Schaub 2002, p. 484). Yet, Bamberg Team 2302 was the most successful in terms of the quality of their design outcome. In a personal communication to me, Joachim Stempfle indicated that Bamberg Team 2302 experienced disagreements and challenges of ideas which nonetheless lead to careful analyses and selection of a design idea.

Let us examine, using LSA, the formation of a coherent design concept in the design conversations of these groups with these prior findings in mind. The graphs shown in Figs. 3.3 to 3.6 show the average semantic coherence between utterances as a function of distance between utterance boundaries in the design team conversations. The dots indicate the value of the semantic coherence; the stars indicate a polynomial curve fit of the data to help assess the smoothness and regularity of the data points. The plots are intriguing in that they appear fairly regular with a non-zero slope initially, but eventually approach an asymptotic limit. Outside a certain distance between utterances, the coherence 'drops off' and is highly scattered. There is a regularity to the graph of the Delft backpack design team (Fig. 3.3) and the Bamberg Team 2302 (Fig. 3.6) whereas the graphs of Bamberg Team 1102 and 2202 appear more 'scattered' and 'chaotic'. This scattering (inconsistent, non-temporally uniform framing of the design concept) is suggestive of weak aggregation and is worth a close examination.

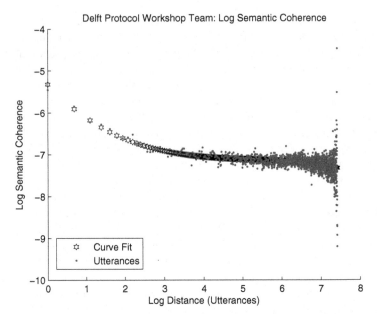

Fig. 3.3 Semantic coherence of the Delft backpack design team conversation

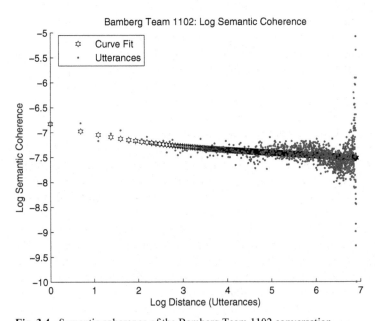

Fig. 3.4 Semantic coherence of the Bamberg Team 1102 conversation

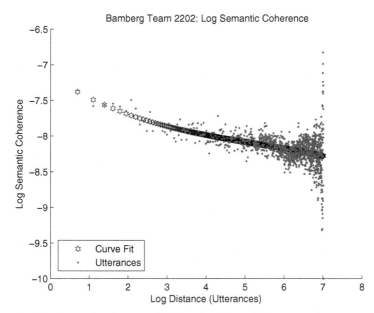

Fig. 3.5   Semantic coherence of the Bamberg Team 2202 conversation

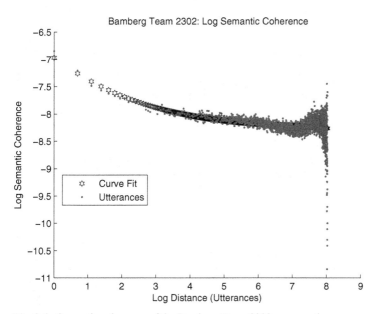

Fig. 3.6   Semantic coherence of the Bamberg Team 2302 conversation

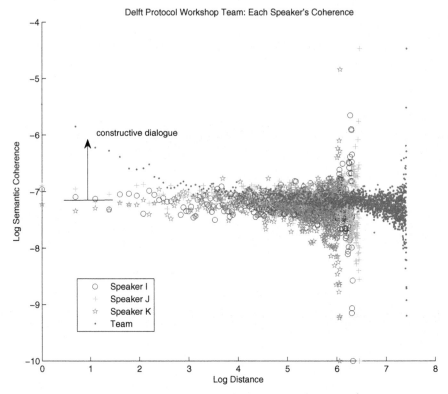

Fig. 3.7 Coherence of each participant for the Delft backpack design team conversation

An intriguing phenomenon occurs when we dissect each speaker from the conversation in order to assess each participant's contribution to the formation of the design concept. Two different patterns emerge. In the pattern that we call the 'constructive' pattern, the design team participants built upon each other's utterances resulting in an increased level of coherence. Even though each speaker's coherence, shown as the open circle, star, and plus sign, is roughly constant, in additive, they increase the coherence of the conversation as indicated in Figs. 3.7 and 3.10. It is as if the contribution of each person to the group's conversation led to a more coherent conversation by the group.

Contrast the graph of Bamberg Team 2302 in Fig. 3.10 to that of Bamberg Team 1102 in Fig. 3.8. Each individual's coherence is scattered within a fairly linear band in Fig. 3.10. Second, there is a general upswing in the group's overall coherence, as shown by the dots, tending to be above the average semantic coherence for each participant for each value of utterance distance. In contrast, the semantic coherence is scattered through a wide band in Fig. 3.8 and there is little evidence of construction. The designers in Bamberg Team 1102 do not use language in a way that frames the design concept. The group's average semantic

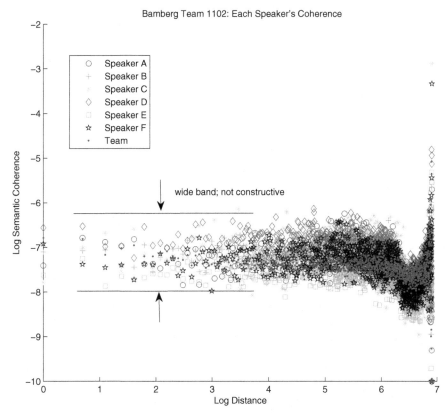

Fig. 3.8 Coherence of each participant for the Bamberg Team 1102 conversation

coherence is erratic. That is, the contribution of each group member to the group's discussion does not increase the semantic coherence of the overall conversation. Their conversation is one in which there was little in the way of building upon each other's expressions of ideas. While each individual designer may have a coherent plan in mind for solving the design problem, there are, apparently, few attempts to reconcile each designer's viewpoints. Bamberg Team 2202 is somewhere in between. There could also be third possibility, destructive conversations in which the overall coherence is lower than the coherence of each participant, but none of these data sets reveal this possibility.

When there are more than two individuals in a team, deriving a representation of the team's shared language that frames the design concept involves intersecting each member's own frame for the design concept. Depending on the level of associative patterns between each individual's utterances, these conversational coherence maps contain semantic meaning (words) that are both explicitly articulated by all the members and tacit meaning that is imputed by LSA from the major associative patterns. We should take care, however, to suggest that the associative

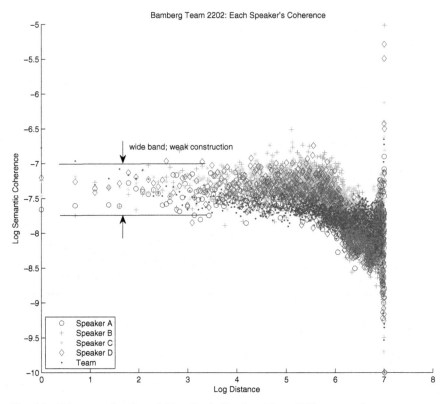

Fig. 3.9 Coherence of each participant for the Bamberg Team 2202 conversation

patterns mean that each participant's semantics is a union of the group's semantics. The union is the frame, the aggregation of their semantics. Rather, the semantics of each participant contains some intersection with the group's semantics. What this means is that the semantics contained in the statements by one participant must be shared by at least one other participant. The degree to which this is shared is a measurement of the strength of coherence of the group's framing.

The data from these studies confirms that the semantic coherence of successful and non-successful teams differs, under equivalent conditions set by the design brief, available technology, and experimental context. Given the attention of LSA on associative patterns of words, the data suggests that, on average, there is greater sharing of semantics in the successful groups over the non-successful groups. The absurd conclusion to draw is that the teams should just repeat each other. Rather, we would expect that the successful group's conversation contains more and different semantic concepts, a linguistic phenomenon noted by Mabogunje (1997), than the non-successful group's conversation. We might also conjecture that the successful group's conversation connects semantic concepts towards an accumulated concept. This is the linguistic pattern that we will take up in the next chapter.

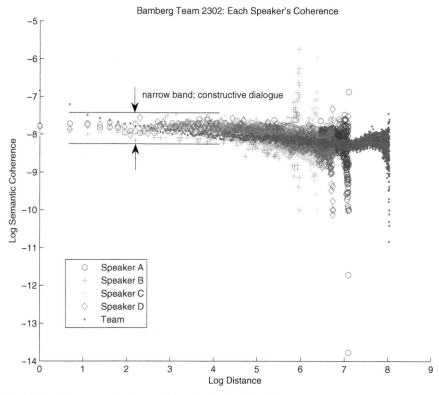

Fig. 3.10 Coherence of each participant for the Bamberg Team 2302 conversation

Summary

These results provide an informational basis for the performative effect of aggregation. Perhaps it is somewhat controversial to suggest that the utterance of the words effected the social cognition of shared understanding. Can it be claimed that the words produced a cognitive effect, that is, that the words preceded shared understanding? Or, should it be claimed that the cognitive effect of shared understanding preceded the public display of the words and that shared understanding causes the display of a semantic similarity between words? What I would find problematic with the latter claim is the presumption, then, that a design concept existed pre-computed in a designer's mind. In the situation of group design, this is unlikely as it is the group, though communication in natural language, which forms the design concept in real-time. In the situation of the designer working alone, a conversation may not take place with other people, but that should not preclude language from playing a role. The Sapir–Whorf hypothesis claims that how we understand and behave is largely codified in the patterns of our languages.

The controversial hypothesis of "linguistic determinism" states that the language available to humans shapes our thoughts. That language allow humans to have thoughts about objects which do not yet exist is not likely to be surprising; but, impoverished conceptualization capacity with language may inhibit humans from generating linguistic expressiveness that could be mapped, enacted and inscribed into a designed work.

Aggregation is not a process unique to the language of design. Cells aggregate to form tissue and then the next level – organs. In the development of cancer cells, for example, tiny cancer cells can aggregate and form their own networks of blood vessels – a process called angiogenesis, which redirects essential nutrients in the blood away from vital organs to the tumor itself. In early stages of cell development after an ovum is fertilized by a sperm, inherent programming in the cell is thought to dictate which cells aggregate to form neural tubing, the spine and other organs. This all happens within the first trimester of pregnancy. Of course, cell aggregation can go right or wrong, leading either to healthy cells and organs, or cancerous cells. Understanding the internal biological mechanisms, external influences, and genetic programming underlying cell aggregation is an exciting area of research.

Likewise, in design, we have seen examples of both felicitous and infelicitous aggregation in design texts. If the end-goal of designers' informal design documentation and conversations is to pool their resources to negotiate different design perspectives and specialties to produce a designed work, then the challenge is to model how language use in design frames a coherent design concept.

Linguistic aggregation is thus thought to play an important role in the process of designing. The rate and level of aggregation, such as with those described in the above examples, vary. The aggregation kinetics shows that the performative operator of aggregation blends ideas-as-words to form larger, multilayered semantic aggregations of design concepts. Latent semantic analysis makes inferential links to semantic meaning through latent, distributed links between semantic content that is contained elsewhere in the design corpus. The semantic coupling through major associative patterns gathers a base set of materials (words) to produce a framing of those words. It is from this frame that structural realizations of the design concept are enacted linguistically. While LSA is useful for analyzing large, terminological trends, we need a more fine-grained approach to explore the actual semantic content in-depth and the mechanism by which ideas-as-words (what we will define as lexicalized concepts) build up to the materiality of the design concept. We take up these problems in the next chapter.

4 Accumulation

> *If a performative provisionally succeeds – then it is not because*
> *an intention successfully governs the action for speech, but only*
> *because that action echoes prior actions, and accumulates the*
> *force of authority through the repetition or citation of a prior*
> *and authoritative set of practices.*

Judith Butler, *Excitable Speech*, p. 51

Building Up to the Designed Work

Accumulation, it could be said, is a reconstruction of the past in a way that suits the current situation. As the epigraph by Judith Butler relays, bringing back to the present time a discourse that recalls or reconstructs a language that was part of the narration brings the authority of the prior discourse as the condition through which it is possible to construct a new narrative. It is this idea of citation that Butler uses to argue how bodies become gendered through the repeated performance of gender. One common mistake that many theorists make of Butler's performative theory of gender is that any gender performance is a 'willful' production of gender. For Butler, gender is not a willful performance but rather a performance which draws on a complex of citational practices sustained through the production and repetition of public, external bodily acts. Performances obtain their power to produce the subject only to the extent that performances are reiterated and sustained across time. Performativity as formulated by Butler derives from the idea that linguistic practices are governed by structures that regulate the realization (i.e., the act of speaking or writing) and recognition (i.e., by peers) of the language. The realization of gender, like the realization of language, follows a set of conditions that govern how language can be structured to accomplish specific functions.

The idea that a lexico-grammatical framework constrains the features available to the user of a language in realizing language is found in the theory of Systemic-Functional Linguistics (SFL) by Michael Halliday (1976), one of the most influential theories of language. According to Halliday, the central focus of language is to do something; the theory of SFL addresses what language does and how it is done. The clause, the most basic lexico-grammatical unit, is realized by three semantic meta-functions of language:

1. ideational: to represent ideas
2. interpersonal: to function as a medium of exchange between people
3. textual: to organize, structure, and hold itself together

A. Dong, *The Language of Design*,
© Springer 2009

Central to the theory of SFL is the network of choices which constrain the user of a language. This network is an inheritance network which means that choices at each level constrain those in the subsequent levels. The choice as to which path through the network is taken in order to achieve one of the semantic meta-functions of language will be determined by the textual structure of the text taken in its entirety and on the social context within which the text is produced, e.g., formal technical writing or informal conversation. The constraints imposed by the structure of a grammar yield the potential to analyze how the structural consequences of that choice relate to how language is utilized as a tool for making meaning.

One of the valuable points to take from Halliday's SFL theory is the placement of the textual structure alongside the social context, drawing attention to the power of the social context to constrain the choices made by the speaker. The inheritance network becomes a framework to register the subtleties of social context, making visible the 'invisible hand' that regulates the set of semantic and grammatical choices made by the speaker in expressing experience through language[1]. The consequence of Halliday's thesis on the centrality of the function of language is that meaning and its linguistic expression is interdependent with linguistic features in terms of the functions they serve in the making of meaning.

The implications toward the language of design of Butler's concept of citation in performativity and the lexico-grammatical framework of SFL are twofold. First, both the way that a concept is expressed by language and the semantics of a lexicalized concept through a language is constrained by expectations as to how the language can be performed. These constraints affect the possible space of design concepts. The hierarchical construction of a linguistic expression therefore maps itself onto structural consequences in the design concept itself as an effect of the choices. The analytic task is to follow the tangle of the citations which result in the accumulation of the materiality of the design concept. Thus, the concern is to turn our attention toward the formation of the design concept as the effect of linguistic practices of citation.

That a designed work that does not yet exist, and may never exist, could be embodied by language is not an unusual concept. Psychologists, beginning with Vygotsky, have researched extensively on this question of how language can refer to or index an object that is not (yet) spatiotemporally present in the context of understanding human psychological development. Vygotsky described these types of situations as cases of "linguistic decontextualization" because language is not necessarily bound to an object nor does the language refer to an object that is spatiotemporally present. Vygotsky was interested in how and when children develop the capacity to unbound relationships between an object and the language used to refer to an object. While the nuances of how language operates contextually independent of objects that the language refers to, Vygotsky's interest in the decontextualized functioning of human language points in the direction of our interest in

---

[1] There are intense debates in the linguistics community on linguistic determinism and linguistic relativism in relation to Systemic-Functional Linguistics. This is not a debate that can be entered in within the scope of this book, but I would be remiss not to acknowledge the controversy.

how it is that language mediates the construction of a concept, a designed work, that does not yet exist. And, while Vygotsky was mostly interested in how children develop an understanding of word meanings and how speech imposes a certain meaning potential on situations, we would like to delve further into how it is that language is transformed into an object.

We need to begin to map out where it is that the structural forms of language intersect with the material construction of the designed work. That is, how is the designed work built up from language? What is at stake in this analysis is how language constructs a materiality for the design concept through words as 'building materials' and semantic links as 'nails', 'glue' and 'mortar'.

This question differs from the issues surrounding how language is used to negotiate a shared understanding and how situations are continually modified by language. The issues involved in addressing this type of question have to do with aspects of common ground and intersubjectivity. The speakers in such a situation must have an agreed upon a foundation for the context of the communication. In contrast to these problems, language in design is being used to produce a context which does not (yet) exist. The context is represented through linguistic units and visual representations. In these situations, I am in accord with Vygotsky's notion of decontextualized meanings since the 'real' context is generally absent and any meaning that exists is predominantly linguistically generated. We have to imagine the situation in which a designer or designers are talking about a designed work with only low fidelity visual representations of the designed work such as sketches and simple models. So when designer John in the Delft Protocols Workshop says, "and then y' just had clip points plastic clip points that just sort of snap around the frame" (t 671), there is no clip point or frame that the language is necessarily referring to as these objects do not yet exist. Or, we have to imagine the situation in which we are reading about a designed work; the decontextualized link to the designed work is produced through language. I can read about the design of a nano-caliper, but it is unlikely that I am actually looking at one.

Thus, language produces the object and is the object referent. The challenge then is to examine, in these situations characterized by a general lack of contextualization and rudimentary object referents, the mechanics of language producing an object definition. My position is that a starting point for understanding the relationship between linguistic representation and object production is to account for the object's existence through the object's explicit construction by decontextualized linguistic meanings. Specifically, how is language instrumental to the construction of the object through its decontextualized meanings? To do this mapping, we will examine language use in the process known as concept generation or conceptual design.

The generation of concepts plays a critical role in the early stages of design and is often considered the 'start' of designing. In practice, concept formation occurs throughout the design process. To be clear, at least two interpretations of concept formation exist. One is the notion of a concept as patterns in phenomenon. In this interpretation, concept formation in design deals with recognizing emergent patterns (Gero 1998) or displacing existing concepts (Schön 1983) when prior ways

of framing phenomena no longer accord with current perceptions or interpretations. The other notion of concept formation in design (and often termed conceptual design) deals with the construction of ideas (e.g., structural embodiments,
functional systems) for designed works. Methods of concept formation such as
brainstorming, frame of reference shifting, synectics and computational techniques
such as evolutionary programming and curious systems make systematic and repeatable what is often considered, though not necessarily factually so (Shah et al.
2005), a creative, non-systematic process. It is this notion of concept generation
that we deal with.

As I argued above, aggregation is not a sufficient linguistic resource to account
for the transformation of representations into a designed work. To develop a designed work by taking in diverse elements and making them into a whole also
requires coming up with a new representation, not just absorbing and framing
experiential and situational elements. In the previous chapter, we dealt with the
concept of aggregation as associated with the collection of experiences and ideas
in such a way that a coherent frame about the designed work is possible. The question we will deal with in this chapter is how language is implicated in the linguistic construction of a new designed work.

We need to add another performative operator to our stable. We recall from the
previous chapter that voices which become aggregated in design text bring with
them prior knowledge and subjective perspectives to a design situation. Language
and the meaning of words encode authoritative knowledge and subjectivities. As
with connectionist theories of the mind, we can conceive of these words in terms of
a network of interconnected units rather than as symbolic processing units. New
design representations emerge from connections across these words rather than
through a prescribed reasoning process. A consequence of this connectionist view
is to regard the words not as variables for symbolic processing but rather as units in
states of connectedness. Connections occur through interrelations of lexicalized
concepts which are partially determined by the semantics and the grammar of language. Each word transmits partial information about a designed work. Through
the production of intersubjectivities between the text and the reader, the lexicalized
concepts may become shorter or less explicit because the text can rely on prior
understandings generated by the prior words and clauses in the text. The accumulation of the words as lexicalized concepts facilitates the 'building up' of words and
clauses into fully-developed concepts. The claim is that accumulation, in the sense
of building up, is one way in which the semantic structures of language scaffold in
order to support the construction of a design concept as a new representation.

This claim is supported by theories in discourse analysis which states that if
two words or concepts are thought about in the same way by a group of people,
then the words will be used systematically in the same way in the conversation of
those people (Lakoff and Johnson 1980). The idea is that the lexicalized concepts
transmit partial information about design concepts and, in doing so, accumulate
the ideas represented as lexicalized concepts into fully-developed concepts.

The performative effect language achieves is to accumulate the material possibilities of the designed work linguistically. This space of possibilities has been

framed through aggregation. The conditions of symbolic expression are then determined through the textual meta-function of language.

Suppose that the following conversation takes place between two roboticists designing robots which traverse a crowded room to serve drinks.

Roboticist 1: We need a different design for the *vision system* for location tracking. The problem is that when people walk in front of the eye, the robot cannot *detect* the other robots and may wander aimlessly. So sometimes you get two robots moving too close to one another or they're far apart and people have to walk toward the robot to get the drink.

Roboticist 2: There's been discussion that we should move towards installing an external *sensor* rather than relying on an onboard *detector*.

Roboticist 1: That's an interesting idea. Like *satellite navigation* with *GPS* in cars.

Roboticist 2: Right. Except perhaps we would use RFID or Wi-Fi triangulation.

Roboticist 1: So we would have an *external controller* to help the robots navigate and just use a simple collision *detection sensor* on the robot to prevent it from hitting objects.

To a trained roboticist with mechatronics design experience, it would be clear that these designers would likely continue towards designing the vision system and the associated software code to control the moving robots. Linguistically, what holds this conversation together is that each of the italicized phrases (elemental lexicalized concepts) relate to a more general (abstract) concept of electronic devices that can receive (detect) an external electromagnetic stimulus. The words *sensor* and *detector* are synonymous. In addition, if we could generate a data structure that accumulates the semantic links between the following elemental lexicalized concepts, *sensor*, *detector*, *RFID* and *Wi-Fi triangulation*, we could derive a general indication about the design concept these roboticists are considering. Though in a rather contrived way for the purposes of illustration, the accumulation of these lexicalized concepts scaffolds the design concept of an *external controller*. The dialogue suggests that the external controller replaces the onboard computer vision system which will be limited to collision detection rather than location tracking.

Accumulation is enacted by referring to lexicalized concepts and by connecting lexicalized concepts through propositions that are operating at different levels of abstraction. Accumulation is depicted diagrammatically in Fig. 4.1. Various voices (the large circles) exist and what the voices carry is partially reflected in the lexicalized concepts (indicated by the horizontal lines with a different line type for each clause in the text). Aggregation produces a frame which contains a collection of lexicalized concepts (the bold-faced circle). However, a collection of words in and of themselves is not sufficient evidence of concept formation; the words (lexicalized concepts) need to be accumulated (shown in the dashed line box to indicate the contribution of voices as lexicalized concepts). The hypothesis is that the constructive mechanism for concept formation is the accumulation of lexicalized concepts. The accumulation is enacted by connecting lexicalized concepts directly (repetition) and at different levels of abstraction.

Thus, the linguistic features by which the accumulation would be evidenced are repetition, anaphora, and hypernyms, a lexical concept that is a generic class of

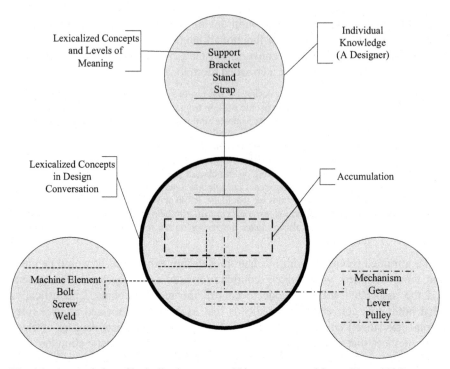

Fig. 4.1 Accumulation of lexicalized concepts within an aggregated frame (Dong 2006)

concepts. Each of these linguistic features could be distinguished as types of semantic links between lexicalized concepts. By analyzing language use in design
for the type of semantic links, accumulation could then be described through a data
structure comprised in the semantic links between lexicalized concepts.

Lexical Chain Analysis

Language Theory

A lexicalized concept is a concept (idea) which has been expressed as a word in
the vocabulary of a given language. A concept can be lexicalized by more than
one word. The underlying assumption of the language theory is that a design concept can be represented by the set of word forms, that is, a set of lexicalized concepts. At the level of granularity of our analysis, lexicalized concepts are not the
type of fully-developed design concepts that would found in Pugh charts. The
lexicalized concepts in our analysis are chunks of knowledge that, when accumulated, may form a fully-developed design concept. The interest is to analyze the
epistemology of concept formation by examining semantic features of the words
in design text to account for the accumulation of lexicalized concepts.

Fig. 4.2 Conceptual model of
a lexical chain

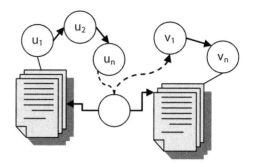

A lexical chain is a sequence of semantically related words in text (or conversation). Lexical chain analysis (LCA) arises from the semantic connections between words which are typically derived from large lexical databases such as WordNet. In computational linguistics, LCA is used, among many applications in text analysis, to derive text paths, analyze the cohesiveness of language use, and to summarize text. Lexical chains capture the cohesive portions of the text and have been shown to mirror the discourse structure of the text (Morris and Hirst 1991). The key to generating the lexical chains lies in the selection of the rules for connecting the lexicalized concepts.

Figure 4.2 illustrates the conceptual model of a lexical chain where u and v symbolize lexicalized concepts (words). The unlabeled node connects the u and v chains, and their associated documents, through a semantic relation even though a direct relation may not have been possible via u_1 and v_3, for example.

Computationally, a lexical chain is generated by using an electronic lexical database containing lexicalized concepts, their gloss and semantic links. WordNet (Fellbaum 1998) is a lexical system which organizes English nouns, verbs, adjectives, and adverbs into lexicalized concepts connected by semantic links. The intent of WordNet was to represent lexicalized concepts in a way that would enable researchers to study the psychology of how humans think about concepts, make connections between and among them, and use context to ascertain the appropriate sense of a lexicalized concept.

Before describing the algorithm for creating lexical chains, we need to define several terms:

1. Gloss: the definition of a lexical concept
2. Sense: The idea that is intended by a lexical concept
3. Synset: a set of one or more synonyms
4. Hypernym: lexical concept which is a generic class of concepts
5. Hyponym: lexical concept which is a member of a class of concepts
6. Meronym: lexical concept which designates a concept as a constituent component of another class

Let us take a concrete example. According to WordNet, the gloss of the lexical concept "pulley" is "a simple machine consisting of a wheel with a groove in which a rope can run to change the direction or point of application of a force

Fig. 4.3 Semantic relations
between lexicalized concepts

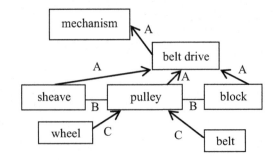

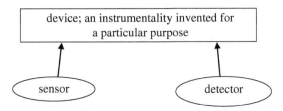

Fig. 4.4 Hypernym (*upward*)
relation between two words –
The concepts "sensor" and
"detector" are a kind of
"device"

applied to the rope". The lexical concepts "wheel" and "belt" are a meronym of the
concept "pulley", both of which are constituent components of a pulley. A "pulley"
is also known as a "sheave" or "block" so they would be part of the synset of the
lexical concept "pulley" at the same semantic level. The concept "belt drive" is
a hypernym of "pulley" and the concept of a "mechanism" would be a hypernym
of "belt drive". The reverse semantic links would be hyponyms. These relationships are illustrated in Fig. 4.3 where A stands for hypernym, B for a synonym,
and C for a meronym.

Examples of the upward (Fig. 4.4), horizontal (Fig. 4.5), part-of (Fig. 4.6), and
downward (Fig. 4.7) relationships for sample lexicalized concepts are illustrated
below.

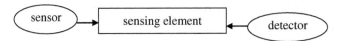

Fig. 4.5 Synonym (horizontal) – A "sensor" is synonymous with a "detector" related by the
concept "sensing element"

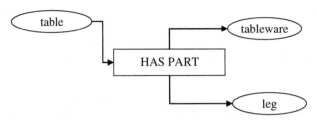

Fig. 4.6 Meronym (part-of) relation between words – A "leg" and "tableware" are a part of a "table"

Fig. 4.7 Hyponym (*downward*) relation between words – A "head-quarters" is a more specific concept of both "office" and "business estab-lishment"

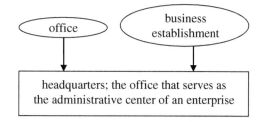

One way to empirically locate accumulation is to analyze the conversation of designers engaged in conceptual design, the stage of design during which initial design concepts (prototypes) are realized. Concept formation in group design is an apt experimental condition for uncovering the performative operator of accumulation. In this situation, individual designers aggregate their expertise to produce a coherent basis from which to realize a new concept. Accumulation in their discourse is the process of adding new knowledge by each succeeding utterance. An 'accumulating' discourse is one that is developed more by a specialized, group specific exchange and sharing of meanings than it is by a set of meanings that might be abstracted or abstracting. Schön describes a type of accumulating discourse as one that is "elliptical and inscrutable to outsiders" (1983, p. 81), and that language is party to constructing "design worlds" (1988) from elemental concepts. When a concept is lexicalized, the semantics of the lexicalized concepts will necessarily impose some structuring on the further possible choice of concepts to be lexicalized. Elementary design concepts could be ideated by simple lexicalized concepts (e.g., gear, backpack). More developed and fully-formed design concepts are accumulated from the elemental ideas.

Computational Implementation

A standard method for constructing lexical chains, such as for text summarization (Silber and McCoy 2002), is to extract the set of lexicalized concepts from the document set and then iteratively add lexicalized concepts to a lexical chain when a semantic link exists between a lexicalized concept and lexicalized concepts in an existing chain. However, our intent is to examine if concept formation could be characterized by the accumulation of knowledge represented through lexicalized concepts through their various levels of meaning. Because we argue that this process happens 'bottom-up', we do not wish to look 'top down' starting from known concepts, but rather to look at concept formation through ongoing linguistic expression and accumulation of lexicalized concepts. Thus, our method proceeds consecutively through the utterances. A lexical chain forms and grows as the result of the accumulation of lexicalized concepts through logically consistent semantic links. A chain breaks when the semantic links no longer support the accumulation of lexicalized concepts.

As a consequence of the interest in examining the building up (accumulation) of words leading to concept formation, assumptions need to be made about the size of the frame (the number of clauses in a text) within which we search for lexical chains. These assumptions have computational complexity consequences on the lexical chain algorithm rather than on the language theory itself.

Assumptions

1. Lexical chain links can only be established between adjacent clauses.

Fig. 4.8   Pattern of links for assumption 1

2. Lexical chain links can be established between a clause and any prior clause.

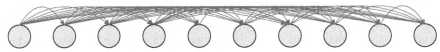

Fig. 4.9   Pattern of links for assumption 2

3. Lexical chain links can be established during a window in a frame for which a set of clauses is semantically close.

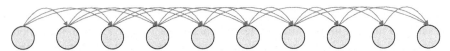

Fig. 4.10   Pattern of links for assumption 3

Assumption 2 leads to an O(n!) algorithm and is thus not computationally tractable. Thus, we do not proceed with this assumption. However, Assumption 2 is equivalent to connecting individual lexical chains through mutual lexicalized concepts even if the chain were broken during the conversation. Assumption 1 is subsumed by Assumption 3 and is redundant. Thus, we proceed under the assumption that a schematic representation of lexicalized concepts is displayed within a small window of semantic consistency; concept formation is enacted by an accumulation of these lexicalized concepts.

The algorithm for constructing a lexical chain proceeds as follows.

1. Filter for the most significant words or phrases in the text. Generally, content-bearing nouns are filtered and only the noun senses are examined.
2. For each word, find the word/phrase in WordNet. If the word/phrase is not found, add the word/phrase to WordNet with a definition (gloss) and link with appropriate synsets.
3. For each clause, select a set of content-bearing lexicalized concepts **U**:

   a. For each u in the selected set of lexicalized concepts **U**, within a window w, create a set of candidate linked lexicalized concepts **V**.

      i. For each candidate v, create a lexical chain between u and v using the following rules:

         (1) The synset(u) relates to a common word x of synset(v) in one Word-Net classification direction upward, downward or horizontal.
         (2) u and v co-occur or repeat within the window w
         (3) v is a meronym of u or vice-versa

The challenge becomes one of finding the optimal size of the window w within which to locate the lexical chains. The window size can be found by segmenting the text into thematically similar chunks where one is likely to find the densest cluster of links. One could manually segment the text by postulating when breaks in the thematic coherence occur. To find the window of thematic coherence, we assumed that the topical focus of the text over sections (such as a paragraph or a chapter) is generally coherent, but that clauses further away from each other would be less likely to be thematically similar. This is a similar to a method developed to examine thematic coherence in written text (Foltz et al. 1998) or as done in the previous chapter. Then, there should exist a mostly ordered relation between semantic similarity and distance between clauses. This ordered relation can be revealed by examining the coherence between any two clauses as a function of the 'distance' between the clauses. That is, instead of calculating the coherence between adjacent clauses, calculate the average coherence between clauses which are one clause away, the average coherence between clauses which are two clauses away, and so forth, to expose the structuring of language over the entire text. This is the same process used to calculate semantic coherence for aggregation in spoken discourse in Chap. 3.

The decay of semantic consistency is applied to segment text for a computationally tractable lexical chain analysis. Coherence is defined per the standard definition. The coherence (cosine similarity) between any two clauses represented by the vectors d_q and d_{q+w} is the dot product of the clause vectors normalized by the product of their norm.

$$\cos\left(d_q, d_{q+w}\right) = \frac{d_q \cdot d_{q+w}}{\|d_q\|\|d_{q+w}\|} \tag{4.1}$$

We then find the best polynomial curve fit f(w) of semantic coherence as a function of clause distance where w is the distance (window) between clauses and f(w) is the value of the semantic coherence. The point on the curve which is half-way between the maximum coherence (the average coherence between immediately adjacent clauses) and the asymptotic limit of coherence is defined as the size of the window of a thematically coherent segment of the text. The asymptotic limit of coherence is defined by the point on the curve fit where the slope of the curve fit is nearest to zero. That is, the location of the asymptotic limit w_a is defined by a real-valued minimum:

$$w_a = \mathrm{Re}\left\{ \min \left\| \frac{\mathrm{d}f(w)}{\mathrm{d}w} \right\| \right\}$$

$$0 \leq w_a \leq w_m$$

where w_m is the total number of clauses.

A bounded minimization search on the derivative of the curve fit locates w_a.[2]

In practice, the choice of the window size w is arbitrary – the higher the value of w, the longer the candidate lexical chains and the higher the number of semantic links. The value of w only affects the speed of the lexical chain analysis, but not the validity of the chains found because the chains can be connected through mutual lexicalized concepts. Thus, if a chain breaks because no semantic links can be found between lexicalized concepts within a window, the chain could nonetheless be joined up with another chain if the two chains share mutual lexicalized concepts. The smaller chains thus join to form a longer accumulated chain of lexicalized concepts. However, a window size of one may miss chains when insufficient content is generated in each utterance. It is recommended that the minimum window size be 2. The automated calculation of the window of thematic coherence enables a fully-automated analysis.

Sample Study

This sample study comes from empirical work on analyzing the formation of design concepts using the semantic relations of words in conversations (Dong 2006). Once again, we will use the same data sets as in our examples of latent semantic

---

[2] While this process sounds complex, it is rather straightforward to implement in MATLAB®. Use polyfit to compute the polynomial curve fit of the semantic coherence and polyder to compute its derivative. With the derivative, you can locate the asymptotic limit using fminbnd. After calculating the midpoint between the asymptotic limit and the maximum coherence, use this value to solve for the roots (using the roots function!) of the curve fit which equal this value. Use the smallest real-valued root as the window size.

analysis, the 1994 Delft Protocols Workshop and the Bamberg Study. For brevity, we will use the notation 1102/D, for example, to refer to Participant D in Bamberg Team 1102. As these data sets come from design conversations, a clause is an utterance and the text is the entire conversation.

During the WordNet analysis, only the noun category for all senses of a lexicalized concept was searched. While word sense disambiguation is a general issue for computational linguistics, a reading of the transcripts indicates that the designers tended to stick with a single sense of a word within the window of thematic coherence. For the thematic coherence analyses, we can process the word by document matrix using latent semantic analysis following the standard procedure (such as retaining dimensions 2 to 101 for the k-reduced matrix) as described in the previous chapter.

The window of coherent conversations for each team is calculated using a third-order curve fit according to Eq. 4.2. The values of the optimal window sizes are summarized in Table 4.1.

Based on these window sizes, we can calculate the accumulation of lexicalized concepts by the number and type of semantic links. The values reported in Table 4.2 to Table 4.5 quantify the frequency of occurrence and the types of semantic links that inter-relate the lexicalized concepts in the respective conversations. In addition to counting the horizontal (synonym), hypernym (a lexical concept which is a generic class of concepts), hyponym (a lexical concept which is a member of a class of concepts) and meronym (a lexical concept which designates a concept as a constituent component of another class) relations, we counted the number of lexicalized concepts generated (a lexicalized concept which has no prior link) and the repetition of lexicalized concepts.

In total and as a percentage of semantic links, the team members repeated words most often (except for Bamberg 2202), which was probably necessary to keep their conversations lexically cohesive. More interesting, though, were the high number and percentage of hypernym relations relative to the other types of semantic links, even in comparison to repetition. If we describe concept formation as the accumulation of lexicalized concepts, then the linguistic pattern of lexicalized concepts connected through higher levels of abstraction (i.e., hypernyms) is a principal pattern of accumulation. For each of the teams, there was at least one person who exhibited higher numbers of hypernym relations in the semantic links relative to the others; we could conjecture that this role is crucial in group design. For the Delft backpack design team, the high number of hypernym relations relative to the others is especially pronounced for John. This result is to be expected given John's known qualitative profile. Aside from 2302/A, the difference in the number of hypernyms cannot be accounted for solely by the number of utterances by each team member. There is not an appreciable difference in the number of utterances by each team member that could account for the significant differences in the number of hypernym relations. For example, 2202/C spoke 2% less often than 2202/A, but had 47% more hypernym relations; 1102/D spoke 1% less often than 1102/A but had 51% more hypernym relations.

Table 4.1   Windows of coherent conversation

Team	Window of Coherence (Utterance Distance)
Delft	4
Bamberg 1102	3
Bamberg 2202	6
Bamberg 2302	10

Table 4.2   Number of chain links Delft backpack design team (w = 4)

Generation		Horizontal		Hypernym		Meronym		Repetition	
I	43	I	29	I	162	I	10	I	241
J	62	J	34	J	323	J	9	J	371
J	41	K	17	K	178	K	5	K	221
Ave.	49	–	27	–	221	–	8	–	278

Table 4.3   Number of chain links Bamberg 1102 Team (w = 3)

Generation		Horizontal		Hypernym		Meronym		Repetition	
A	12	A	1	A	36	A	0	A	50
B	20	B	9	B	47	B	1	B	112
C	18	C	1	C	37	C	1	C	50
D	29	D	4	D	74	D	1	D	118
E	20	E	5	E	24	E	2	E	51
F	16	F	4	F	25	F	1	F	65
Ave.	19	–	4	–	41	–	1	–	74

Table 4.4   Number of chain links Bamberg 2202 Team (w = 6)

Generation		Horizontal		Hypernym		Meronym		Repetition	
A	55	A	123	A	278	A	82	A	232
B	25	B	46	B	162	B	32	B	118
C	81	C	219	C	523	C	176	C	396
D	43	D	114	D	210	D	88	D	211
Ave.	51	–	126	–	293	–	95	–	239

Table 4.5   Number of chain links Bamberg 2302 Team (w = 10)

Generation		Horizontal		Hypernym		Meronym		Repetition	
A	159	A	1994	A	2808	A	868	A	3018
B	50	B	428	B	636	B	217	B	712
C	77	C	877	C	1335	C	477	C	1365
D	143	D	1459	D	2234	D	601	D	2023
Ave.	107	–	1190	–	1753	–	541	–	1780

Two other potential explanations exist for this phenomenon. First, John, 1102/D, 2202/C and 2302/A could be what Sonnenwald (1996) has characterized as "interdisciplinary stars" by their ability to abstract knowledge from others and then to accumulate concepts from others. Second, because experts in a domain reason at more abstract levels (Zeitz 1997), hypernym relations may serve as a proxy for identifying the expert in a design group if one accepts the assumption that one characteristic of experts is that they tend to engage in abstract reasoning more often than non-experts.

Because the lexical chains are constructed by successively chaining utterances expressed by different participants in the team, lexicalized concepts flow through and from each participant, with the language of design accumulating as each successive participant linguistically constructs the design concept. It would be interesting to quantify the flow by participant as this would uncover the accumulator in the discourse. To quantify the flow of concepts, we define the following relationships between u and v:

- A weak relationship (value = 1) if the synset(u) relates to a common word x of synset(v) in one WordNet classification direction, upward, downward or horizontal
- A strong relationship (value = 3) if u and v co-occur or repeat in adjacent utterances or, v is a part of u, or u and v share common parts (meronym)

These relationships will be used to assess the influence of lexicalized concepts between the designers, or what we term the 'strength of ties' between the designers. The strength of ties for the Delft backpack design team (Fig. 4.11) and the Bamberg teams (Figs. 4.12–4.14) show different patterns for the flow of lexicalized concepts.

For each team, there is a clear central participant (indicated by the gray circle) who has the strongest set of ties to the other participants. The strongest pair is highlighted in gray in the adjacent table. John, 1102/D, 2202/C, and 2302/A are central, consistent with their high number of hypernym links within each respective team.

These "interdisciplinary stars" that Sonnenwald named performed a crucial linguistic role, which I would describe as a linguistic accumulator. An accumulating discourse is one that is developed more by a specialized, group specific exchange and sharing of meanings than it is by a set of meanings that might be abstracted or

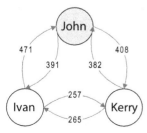

Pair	Strength	Pair	Strength
Ivan→John	471	John→Ivan	391
Ivan→Kerry	257	Kerry→Ivan	265
John→Kerry	408	Kerry→John	382

Fig. 4.11 The strength of the ties for the Delft backpack design team

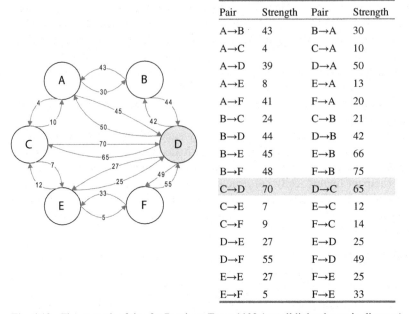

Pair	Strength	Pair	Strength
A→B	43	B→A	30
A→C	4	C→A	10
A→D	39	D→A	50
A→E	8	E→A	13
A→F	41	F→A	20
B→C	24	C→B	21
B→D	44	D→B	42
B→E	45	E→B	66
B→F	48	F→B	75
C→D	70	D→C	65
C→E	7	E→C	12
C→F	9	F→C	14
D→E	27	E→D	25
D→F	55	F→D	49
E→E	27	F→E	25
E→F	5	F→E	33

Fig. 4.12 The strength of ties for Bamberg Team 1102 (not all links shown in diagram)

abstracting. A linguistic accumulator helps to produce such a discourse through a lexico-syntactic behavior that facilitates the accumulation of design content. The manifestation of hypernym relations is one example of such a lexico-syntactic pattern.

John has a strong influence on Ivan and Kerry because concepts flow through and from him; Goldschmidt characterized John's influence as having more "critical moves" (1996). We note also that the strength of ties pointing to John is higher than pointing away from John suggesting the interdependency of John upon Ivan

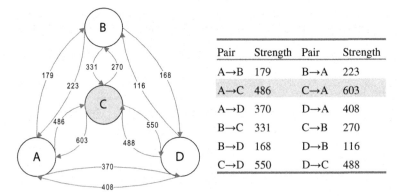

Pair	Strength	Pair	Strength
A→B	179	B→A	223
A→C	486	C→A	603
A→D	370	D→A	408
B→C	331	C→B	270
B→D	168	D→B	116
C→D	550	D→C	488

Fig. 4.13 The strength of ties for Bamberg Team 2202

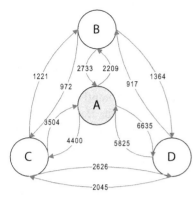

Pair	Strength	Pair	Strength
A→B	2209	B→A	2733
A→C	4400	C→A	3504
A→D	6635	D→A	5825
B→C	972	C→B	1221
B→D	1364	D→B	917
C→D	2626	D→C	2045

Fig. 4.14 The strength of the ties for Bamberg Team 2302

and Kerry. That is, while John is a productive individual idea generator, he also builds upon Ivan's and Kerry's ideas. For example, when John accumulates the discourse, up to a point in the conversation, and names their concept as 'a pack conversion kit that we have to sell them' (t 535), it is known within this group that the 'pack conversion kit' has a particular meaning. The meaning has been generated and encoded into this wording by this group in this design conversation. Each member of the group would, having participated in the conversation, have some shared understanding of the meaning that they have accumulated together. Outsiders (non-group members) would require explanation of what 'a pack conversion kit' is and what it does.

Again, the percentage of contributions to the team's conversation cannot fully account for the stronger strength of ties among some of the team participants. In the case of Bamberg Team 2302, one would expect 2302/A to have much stronger ties than the others because this participant dominated the conversation and, indeed, this is the case. However, 1102/D and 2202/C have a stronger total strength of ties to the other participants despite speaking less often than the most frequent speakers 1102/B and 2202/A, respectively.

Instead, we might characterize these linguistic accumulators as having the ability to connect with and join ideas from other team members. Further, the evidence suggests that this ability is not necessarily related to assertiveness in terms of speaking more often than others. Whereas the other participants state and name concepts, the linguistic accumulators recombine them to effectuate a concept. Thus, although concept formation requires each designer to bridge indirect relations among the concepts stored in each designer's mind, the evidence suggests the need for and existence of a person in the design team who specializes in the linguistic assembly of the concepts. At least one participant is needed to enact the performative operator of accumulation, that is, to be the linguistic accumulator.

Because there were six team members in Bamberg Team 1102, as opposed to four in both Bamberg Team 2202 and 2302, the flow of the conversation may have been impeded due to extra effort expended in coordinating the conversation or a subgroup forming within the larger group, which may account for the differences in the production of hypernym relations. The data suggest that other factors may be operating, however. First, each team member in Bamberg Team 1102, except 1102/C, contributed roughly equivalently to the conversation. Second, and more interestingly, the Bamberg researchers reported that "proceeding in the design task can be labeled from 'chaotic' (group 3) to 'planned' (group 1)" where group 3 is Bamberg Team 2302 and group 1 is Bamberg Team 1102 (Stempfle and Badke-Schaub 2002, p. 484). Third, the number of hypernym links for Bamberg Team 1102 is at least an order of magnitude lower than Bamberg Team 2202 and Bamberg Team 2302. Conversation flow and the number of people do not appear to justifiably account for this discrepancy.

I would conjecture that the planned behavior of Bamberg Team 1102 could account for the higher percentage of repetitions relative to hypernyms when compared to the other Bamberg teams and the Delft backpack design team. That is, their planned design behavior manifests in the linguistic behavior described by Table 4.3; each team member reproduces and repeats lexicalized concepts. Instead, the other teams exhibit differentiated repetition of lexicalized concepts. The differentiated repetition appears as hypernyms that connect concepts that are similar yet differentiated by a level of abstraction. This behavior of differentiated repetition may signal the productivity of language use in design toward the formation of concepts.

Consistent with other similar findings by researchers in human-computer interaction, the participants in each of these studies seemed to play a linguistically definable role. In a study of 10 early-stage software design meetings, Olson and colleagues (1992) found that each participant played a different design role including criteria manager, issue manager, and meeting manager. They concluded that each role allowed the participants to share the cognitive load of the design work. It is unclear whether such roles are always agreed upon ahead of time; certainly, in the studies presented in this chapter, those roles were not pre-negotiated. They happened spontaneously.

What we see in these analyses is that the choice of a particular concept or way of proposing a concept by a designer seems to trigger other related concepts from other designers. When a concept is lexicalized, the semantics of the lexicalized concept will necessarily impose some structuring on the further possible choice of concepts to be lexicalized. One caveat of this method is that the analyses do not make any claim about the type of reasoning or cognitive structures underlying the semantic relationships and the linguistic behavior. That is, we could not claim that a horizontal relationship is analogical reasoning or that hypernymic relations result from conceptual chunking of a functional representation of the device.

But the effect of accumulation is differential; that is, in not all cases does language accumulate. In not all cases is the performative force of accumulation strong. We can see the differential effect of accumulation in another way. Let us separate the constitutive contributions of each team member to the overall coherence of the design conversations, that is, the contribution of each speaker's language to the accumulative force to form a coherent design concept. If language is accumulating to form the design concept, then the accumulated language across all speakers should exhibit overall thematic coherence that is higher than that for each of the constitutive components of the language for the concept. Higher thematic coherence also means that there is more overlap in the language, more accumulation of lexicalized concepts. That is, we would expect additive behavior. The lexicalized concepts augment and build upon one another, and there is a genuine co-construction of the design concept.

We have already seen in Fig. 3.8 (Bamberg Team 1102) and Fig. 3.9 (Bamberg Team 2202), each individual's coherence (non-filled shapes) is scattered within a fairly linear band that also bounds the group's overall thematic coherence, as shown by the solid dots. Each person in the group appears to be saying (contributing) approximately the same concepts. Conversely, as shown in Fig. 3.10 (Bamberg Team 2302), although each speaker's coherence (shown by the non-filled shapes) is roughly constant, in combination (solid dots), they increased the thematic coherence of the conversation. A comparison of Bamberg Team 1102 and Bamberg Team 2302 is shown side-by-side in Fig. 4.15 to highlight the difference. Thus, despite

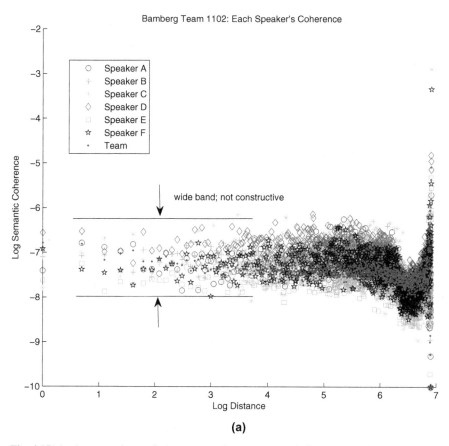

(a)

Fig. 4.15(a) A comparison of the constructive property of the conversation in Bamberg Team 1102

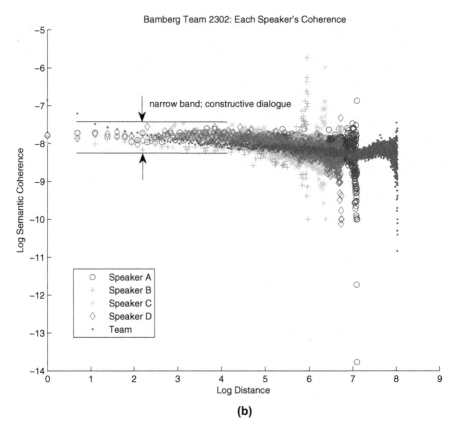

Fig. 4.15(b) A comparison of the constructive property of the conversation in Bamberg Team 2302

Bamberg Team 2302 being labeled chaotic and not following any prescribed design methodology, the statistical patterns of semantic links (e.g., the hypernym relations) illustrate a high level of accumulation of concepts and a pattern of individual to group thematic coherence. This additive thematic coherence allows us an additional means to characterize the performative effect of accumulation. The language of the Delft backpack design team is similarly accumulating, as was shown in Fig. 3.6. These characterizations from the quantitative data are consistent with the qualitative observations.

The phenomenon of linguistic accumulation is slightly easier to visualize when we think of the behavior as a combination of aggregation and accumulation. As language is realized in a design text, the language should refer to the prior practices. Yet, we must also expect that each new text introduces new semantics. This process of citation and production of new lexicalized concepts is what we have seen through lexical chain analysis. Yet, the linguistic pattern of aggregation means that the semantics should progressively cohere toward a semantic frame for the design concept. This behavior implies an internalization of the language. It is possible to map out the kinetics of linguistic accumulation. At each point in time, we calculate the (cosine) similarity of the concept centroid c to each participant's concept centroid c_i. Recall that the concept centroid is the arithmetic mean of the document vectors. This calculation is a measure of the semantic similarity of the framing of the design concept as it is framed by the language of the group and the language of each individual. As we have already seen stark differences between Bamberg Team 1102 and Bamberg Team 2302, we compare their data sets to illustrate the differential kinetics of linguistic accumulation.

Linguistic accumulation in the well-performing Bamberg Team 2302 exhibits strong semantic similarity over time (Fig. 4.16), suggesting a similarity in the ways in which the language of each participant frames the design concept and the way that the group frames the design concept. In contrast, linguistic accumulation of Bamberg Team 1102 is less unified (Fig. 4.17). Person E is shown as not sharing a common frame with the group. This result is consistent with the observation that this team exhibited a lack of common understanding (Stempfle and Badke-Schaub 2002).

We should exercise a bit of caution in extrapolating these kinetic results to any group design situation. Conversational situations such as a design review meeting in which each participant contributes a separate presentation or a planning meeting in which each participant is expected to contribute only to specific aspects of the

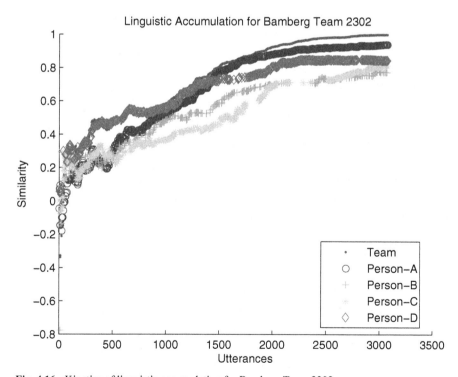

Fig. 4.16 Kinetics of linguistic accumulation for Bamberg Team 2302

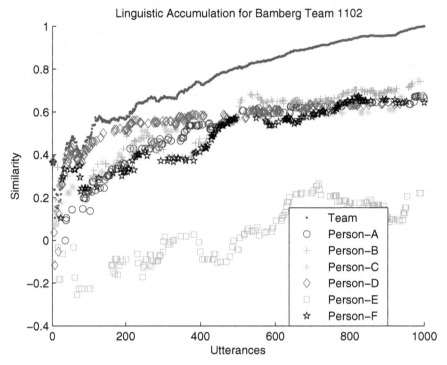

Fig. 4.17 Kinetics of linguistic accumulation for Bamberg Team 1102

conversation may not exhibit this linguistic accumulation effect. That does not mean that these conversations are dysfunctional or that design is not happening. Rather, the participants might be playing specific roles and the time frame for the analysis is too short. In the data sets for the cited studies, there was an expectation that the groups would achieve some sort of consensus on the design concept. As the experimental conditions required that the group form a design concept from scratch, with no prior knowledge of the design problem, it is to be expected that linguistic accumulation *must* occur for a successful design concept to emerge from the conversation. In summary, the data set and the time frame for the analysis of the kinetics of linguistic accumulation must be chosen with care such that there is an expectation that a design concept should have emerged. In industry, a data set consisting of documents and conversations leading up to a stage-gate review would constitute a useful set. Under such a circumstance, if no linguistic accumulation happens or if the kinetics is non-uniform, then we would have good reason to suspect that the performance of design is infelicitous.

It is also interesting to visualize what the accumulated lexicalized concepts were. To create this visual, we can connect each lexical chain through mutual lexicalized concepts and contextualize them to the originating utterance. The connection of the lexicalized chains to their originating utterances allows an interrogation ('reading') of the accumulated language in context. Sample chains from the

Delft backpack design team and Bamberg Team 2302 are shown in Fig. 4.18 and Fig. 4.19, respectively. For illustration purposes, the chosen set of linked utterances for the Delft backpack design team in Fig. 4.18 is interweaved whereas the linked structure of the Bamberg Team 2302 in Fig. 4.19 is linear in time. In practice, the structure of the connected chains is complex and has many possible paths. In the examples provided below, a representative is chosen to expose key utterances in the conversation. In the figures, connections on the top of the circles represent hypernym links whereas the horizontal links represent synonyms or repetitions. A curved arrow denotes the exclusion of interim utterances.

The chains shown in Fig. 4.18 refer to the design objective for the backpack team, attaching a bicyclist's backpack to the rack. What is interesting, as the utterances

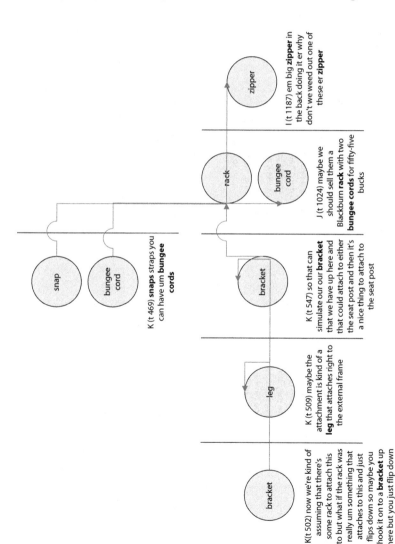

Fig. 4.18 The connected links from the Delft backpack design team (Dong 2006)

show, is that the team attempted to resolve, in parallel, the design concepts for connecting the rack to the bicycle (with various types of brackets) with design concepts for attaching equipment (with bungee cords, zippers, nets, and snaps) to the backpack. In fact, the choice of materials for manufacturing the rack (injection molding, vacuum forming) is directly related to the possible type of attachment concept. Valkenburg (1998) noted that the process of coming to agreement about the materials, manufacturing method, and backpack fastening device lead to a shared understanding about the agreed-upon manufacturing method: injection molding. Although the lexical chains are not sufficient to ascertain whether or not

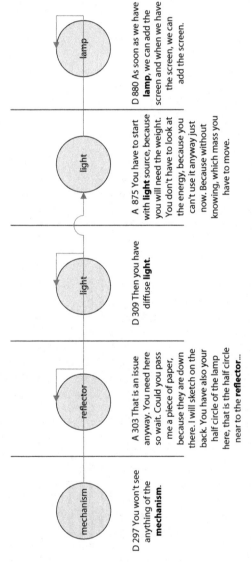

Fig. 4.19 The connected links from Bamberg Team 2302 (Dong 2006)

the team came to a shared understanding, the data calculated in generating the chain usefully indicates the extent to which these concepts were discussed and the accumulation of concepts throughout the design team's conversation.

Figure 4.19 illustrates a set of linked utterances during which Bamberg Team 2302 dealt with the issue of positioning the reflected light that the planetary gear mechanism moves. When 2302/D begins talking about a "mechanism", it makes sense that 2302/A refers to the "reflector", which is what the mechanism will be physically connected to. The reflector will produce a certain type of "light". The light may alternatively be produced by a "lamp".

Human thinking and ability to develop knowledge is heavily constrained by, and made possible through, language. The evidence shown in these sample implementations suggests that language is essential for concept formation. The linguistic behavior to construct a concept is derived through statistical language processing of the types and frequencies of semantic links. The statistical language processing perspective lets us see concept formation by the language of design as a bottom-up accumulation of information and ideas coded in lexicalized concepts. Taking the approach of this and the prior chapter together, we have folded together computational linguistics tools derived from the statistical language processing with the perspective that semantic links are a useful way to model the production of designed work in terms of linguistic abstraction and to establish a new method to examine concept formation in design through computational linguistic analysis.

Summary

The accumulation of language through the connection of lexicalized concepts has been deduced through statistical language processing of patterns of word co-occurrence and semantic links. These analyses illustrated how language served to build up toward design concepts while also functioning in the semantic meta-function of holding itself together as theorized by Michael Halliday. The differential results between the various conversations by the design teams studied would suggest that even where language performs its semantic meta-function of holding itself together, this does not correlate to the production of a design concept. The formation of design concepts was reflected in the particular patterns of accumulated semantic connections between lexicalized concepts. These semantic connections enact the performative power of accumulation in order to produce the design concepts. The design concept is effected from connections across lexicalized concepts. A consequence of this finding is to regard designed works as existing through various states of connectedness of aggregated language, and where variations in states of connectedness could give rise to other designed works. Aggregated language is not just the residue of symbolic processing but rather the designed work in a nascent state of connectedness. The connections that occur through the interrelation of lexicalized concepts accumulate the aggregated language, represented as lexicalized concepts, into fully developed concepts. Lexical chains of the semantic relations enact the performative operator of accumulation.

What we have found from the results in the two preceding chapters is that language provides a substrate for the material that could produce a design concept. Patterns of aggregated and accumulated language seem to predict design outcomes. Going further, the patterns suggest that aggregated and accumulated language seem to coexist if not predicate design, that is, provide a condition of possibility of design. We might conjecture that the ability to generate a design concept is heavily constrained by, and made possible through, language accumulation. In the case of the lexical chain analysis, the choice of a particular concept or way of proposing a concept by a designer seemed to trigger other related concepts from other designers. When a concept is lexicalized, the semantics of the lexicalized concept will necessarily impose some structuring on the further possible choice of concepts to be lexicalized. The linguistic behavior of differentiated repetition manifested as hypernym links appears to impact the productive formation of a design concept. The connections between the semantic links could be interpreted as constituting the accumulation that gives rise to the materiality of the designed work.

5 Appraisal

For the function of language is not to inform but to evoke.

Jacques Lacan, *The Function and Field of Speech and Language in Psychoanalysis*, p. 86

Design and Affect

Design is actioned by desire and intent – the desire to create, the desire to fulfill a need, the desire to please. The desire-intent dyad of design differentiates design from emergent, self-organizing, and swarm intelligence systems. Design text, if it could be temporarily made into a body as Katherine Hayles transforms books and bodies, "is at once a physical object and a space of representation, a body and a message." (Hayles 2002, p. 155) This body can produce and can feel exhilaration, wherein somatic, proprioceptive, and kinesthetic euphoria is dispersed throughout the words. In claiming that design is actioned by desire, what we're really saying is that every activity that happens during design, from (what is observed as) rational decision-making to model-making, are all in the service of affect. What we are saying is that we cannot have any full account of the language of design without thinking about how the language of design 'feels' – that is, declare its dispositions toward the subject. If aggregation and accumulation effect the transformation of representations into a new, coherent concept, then there must exist a way for the language to exclude one representation in favor of another, or to declare one or more concepts as viable but not others, and to interlock emotion with design thinking. In reifying the representations, the language can resist materialization of a concept as much as it can privilege others. Instead of thinking of design as bounded by rationality, design can be imagined as articulated moments of affect in networks of relations with rationality.

One thing is for certain. Without emotions, to borrow the geological metaphor of Marcel Proust cited by Martha Nussbaum (2001, p. vii) from *Remembrance of Things Past*, designing would be an unemotional "plane so flat that even from a good vantage point one could not have discerned an idea sticking up above the ground".

Before explaining how the language of design feels, we should be more careful about what we mean about 'affect', 'feeling', and 'emotion' – all of which are nefarious terms. Sometimes we confound them. The statements, 'I feel tired' and 'I feel happy' are commonly-used linguistic propositions of feelings. Which of these three do we mean when we dealing with 'feeling' in the language of design? I should also be careful to note that I do not mean to discuss why users have emotional responses to designed works, as Donald Norman relates in his book *Emotional Design*.

Brian Massumi clarifies the definition of affect, feeling, and emotions in his introduction to Deleuze and Guattari's *A Thousand Plateaus*. For Deleuze and Guattari, affect is an intensive vibration arising from the body's ability to respond to external stimuli. Feelings are what we subjectively know about the affective state that we are in. What is knowable about affect is based on prior experiences and the capacity for labeling (e.g., linguistic naming) an affective state. Emotions are the publicly displayable and accessible expressions of feelings. Emotions are social in the sense that they are intended to communicate our affective state to others, or to fulfill social obligations, such as crying at a funeral or smiling at a wedding, or to negotiate interpersonal relations.

These definitions somewhat parallel the ones proposed by Sylvin Tomkins (1962); according to his Affect Theory, affect is an innate, biological response to an external stimuli, feeling is the awareness and knowledge of affect, and emotion is affect that has been layered with background experiences. If we remove affect from organisms, if we propose that biology is not required for affect, then we can take affect to be an abstract and unstructured force. "The autonomy of affect is its participation in the virtual. Its autonomy is its openness. Affect is autonomous to the degree to which it escapes confinement in the particular body whose vitality, or potential for interaction, it is." (Massumi 1995, p. 96)

This abstract power can be transmitted from one body to another body, from the design text to the designed work. While we cannot presume that affect can be fully realized through or by language, emotions can be displayed through language, giving us a gauge of the transformative potential of affect in the language of design. The transmission of affect from the text to the designed work is displayed as appraisal in the language of design. Appraisal is way that the language of design marks its attitude toward the experiential elements of the language. Appraisal, as O'Toole (1994) describes, functions as the interpersonal mediator between the design work and the viewer, marking language's attitude toward the work.

In order to prepare for an excavation of emotions (the display of affect) in design, we need to limit the scope of emotions in design as the affective state arising from and/or leading to an evaluative weighing of situations encountered on the goodness of the designed work and design practice. Donald Schön termed these situations and how designers deal with them as reflection-in/on-action. Reflection-in/on-action is their response to affect being transmitted to them by the situation. He argued that rational, codified thinking processes may not adequately describe what actually takes place during design practice. Schön challenged the technical rationality of professions by proposing reflection-in/on-action as a way of thinking that enables professionals to deal with divergent situations, make sense of uncertain situations, and bring understanding to knowledge-in-action. In design practice,

Schön (1983, pp. 76–104) described how reflection allows a designer to navigate a web of possibilities to ascertain "moves". This formulation of reflection-in/on-action is suggestive of Deleuze and Guattari's approach to affect. For them, when a source is expressing an intensity near another body, one intensity is enfolded into another. The responding body resonates with the intensity of the contexts it enfolds. The reflective practitioner is, in this sense, enfolding and responding to the affective resonances of the situation. The emotion displayed is the residue of affect. What emotion is displayed as an effect of the transmission of affect is, however, not a pure expression of the abstract affective force. The expression of emotion is conditioned by background knowledge and the experience of the affect in relation to the potential subjective response by the body. Massumi expresses these attributes in the following way. "An emotion is a subjective content, the socio-linguistic fixing of the quality of an experience which is from that point onward defined as personal." (1995, p. 88)

In other words, we are saying that there are significant internal evaluative and externally conditioning components to emotions. For organisms that can have emotions, external conditions are provided by the context. For humans, the internal evaluative component requires cognition. I do not need to 'think about' feeling tired if my body is indeed physically and physiologically tired from too much exercise or too much work. If I am no longer physically or physiologically tired or hungry, I cannot 'think' my way into tiredness or hunger (try as we might!) when we are simply not. Statements such as 'I feel tired' and 'I feel hungry' do not satisfy what we mean by emotions.

However, I do feel that people, places or events in my life are sources of resonances that make me happy or sad. I can feel happy when I look at the photo of my dog Norton sitting at the edge of Echo Lake in Desolation Wilderness, remembering the first time we went overnight camping. I can feel regret when I look at my grandmother's jade broach that she gave me, lamenting the sorrow of never having met her as she died before I was born. There is, as Martha Nussbaum rightly points out in her book *Upheavals of Thought*, a significant evidence for the view of emotions as eudaimonistic judgments. Nussbaum writes about emotions as cognitive-evaluative judgments involving eudaimonism in the following way:

> The value perceived in the object appears to be of a particular sort. It appears to make reference to the person's own flourishing. The object of the emotion is seen as important for some role it plays in the person's own life. (Nussbaum 2001, p. 30)

When Nussbaum refers to flourishing, she intends to capture the "self-referential element" of the judgment and the "element of general evaluation" of the object that is of importance to the person. Thus, Nussbaum does not mean flourishing just in the sense of fulfillment or self-actualization.

Nussbaum's object-relations cognitive-evaluative view of emotions is closely related to the Ortony–Clare–Collins (OCC) cognitive model of emotions (Ortony et al. 1988). According to the OCC model of the cognitive structure of emotions, emotions arise as a valenced reaction to objects, agents and events. According to the OCC model of emotions, the following would be emotions. I feel proud of my new ecological laundry machine (object), happy because my family is visiting from overseas (event), and pleased about my affectionate dog (an agent). My positive appraisal of the laundry machine is the effect of my attitude towards environmentally-conscious and sustainable technologies and the 'appeal' of a washing machine which exceeds stringent conditions for water and energy consumption. The positive appraisal leads to the feeling or the emotional state of being proud. My positive appraisal of my family visiting from overseas satisfies my goal of staying connected with my family and the desirability of maintaining family relations despite living on the other side of the Pacific Ocean. I feel affection toward my dog because he (usually) satisfies the normative criterion that pets should offer companionship for their guardians[1].

Thus, what we mean by an emotion in the language of design is that the emotion is always intentionally directed, whether at an object, an event or a person. There is a significant element of judgment which is related to its resonance with affect. While the judgment might at times appear so visceral such that we might not even begin to propose a linguistic proposition to explain, account for, or even consciously express the judgment, we are not surprised that the judgment is connected to our thoughts and that the judgment is intentionally directed.

This perspective is consistent with hypotheses put forward in neurobiological and psychological theories of affect and the appraisal theory of emotions. Affect is the neurobiological state incorporating emotion, feelings and other affective phenomena such as mood and temperament. Neuroscientists model affective judgments as a mechanism for organisms to unconditionally and conditionally valuate situations (Burgdorf and Panksepp 2006) as the basis for action. Neurobiologists define the affective processor as "components of the nervous system (conceptual and neurophysiological) involved in appetitive (positive) and aversive (negative) information processing." (Cacioppo and Bernston 1999, p. 133) For neuroscientists, emotions prepare an organism for action because they are an "episode of coordinated brain, autonomic, and behavioral changes that facilitate a response to an external or internal event of significance for the organism." (Davidson et al. 2003) An emotion is a response to an appraised meaning rather than an objective set of qualitatively different and categorically distinct "emotions" such as happiness, sadness, and anger. While not all affective conditions are emotions, affective judgments on the state of the situation, how the situation is believed to come about, and the implications of the situation, all precede emotions, according to the appraisal theory of emotions (Ellsworth and Scherer 2003).

What then of language? Language is vital to the expression of emotions. It is, of course, not the only way that organisms can consciously express emotions. Emotions manifest in various and often ambiguous ways. Physiological expressions of emotions include facial expression (e.g., smiles) and body gestures (e.g., arms crossed) and are associated with a pattern of expression (e.g., a smile is generally

---

[1] Various municipalities in the United States such as San Francisco and West Hollywood have passed so-called "pet guardian ordinances" which changed the notion of "ownership" of pets to companionship.

an expression of happiness) (Ekman and Friesen 1975). Some of these patterns have been shown to be universal with respect to cultures (Ekman et al. 1987). Bio-physiological reactions to emotions include changes to blood pressure, skin conductivity, heart rate, and brain activity, all of which could be measured with appropriate machinery and correlated to various emotions.

The available research in psychology (Pennebaker and King 1999) and linguistics points to systematic approaches to derive reliable linguistic resources that deal with both the representation and the representability of emotions through language. Language enables humans to express their conscious awareness of emotions; therefore, language allows others to assess the subjective feeling component of emotions. There has been detailed research into the language of emotions. Ortony (1987) established an "affective lexicon", a taxonomy that categorizes words about affect into affective, behavioral and cognitive "foci". The criterion for the placement of words in each of the foci is based upon the extent to which the word refers to how one is feeling (affect), acting (behavior), or thinking (cognitive). In linguistics, the means for expressing emotions is called appraisal. It is through the linguistic system of appraisal that language displays toward the judgments that are the basis of emotions in the language of design.

However, perhaps it is naïve to state that language is just expressing emotions. The concept of language versus emotions maintains a mind–body split that might be unintentional and unnecessary for our purposes, which is to discover the ways in which the text filigrees semantic meanings to remove its silence on its positions toward those meanings. That the expression of emotion also has a generative cognitive-emotional effect is being supported by recent research on cognition and emotion. Forgas (2000), for example, claims that positive affect benefits creative activities such as design though designing was not explicitly studied.

Design studies researchers are just beginning to discover how emotions may directly influence design, such as having equivalent effect on ideation during conceptual design as formal reasoning (Solovyova 2003). Negative appraisals are accompanied by a focus on analysis of situational data whereas positive affective content displays a reliance on background experience. My colleagues and I recently completed a study on the influence of the valence of the affective appraisals, as the orientation of the linguistic appraisals, on knowledge integration and generation (Dong et al. 2007). We found that positive affect has a congruency effect with knowledge generation and negative affect hampers/inhibits knowledge generation. What is most interesting is the asymmetric impact of affect on design cognition in relation to knowledge as beliefs or knowledge as objective, verifiable facts. It turns out that knowledge as beliefs is more affect-sensitive, that is, beliefs are more likely than objective knowledge to be sensitive to affect. The occurrence of affective appraisals when the subject of the appraisal is closely tied to the person's beliefs rather than objective knowledge suggests that beliefs are not easily adjusted to be compatible with internal evidence in the form of affect. Our findings are suggesting that the valence of affect exercises regulatory effects consequential to design thinking.

In summary, my perspective is that affect provides the force of will in support of or refusal of aggregation and accumulation. The affect is displayed linguistically in appraisal, the appetitive (positive) and aversive (negative) representations of responses to affective resonances. These appraisals position a layer of attitudinal stance on top of an account of design to propel the way design is practiced and to regulate that which will be produced. The language of design transmits affect to the designed work through the performative operator of appraisal. The linguistic expression of affect "can resonate with and amplify [affect] at the price of making itself functionally redundant. When on the other hand it doubles a sequence of movements in order to add something to it in the way of meaningful progression – in this case a sense of futurity, expectation, an intimation of what comes next in a conventional progression – then it runs counter to and dampens the intensity." (Massumi 1995, p. 86) Design is actioned in the presence of this affective flow that interprets and validates, yielding interventions and appraisals.

Our project in this chapter is to lay out the language of appraisal in design and uncover how affect works through it to regulate that which will be produced. The language of appraisal deals with a text's display of valenced reactions to the design context or situation. Appraisal is proposed as the term to characterize the set of mechanisms by which the language of design reflects and shows dispositions toward the effect of aggregation and accumulation. Appraisal, in transmitting affect to these information processing operations, guides and modulates the materialization of design. What is observed as cognitive enactments of affect are perhaps, rather, a temporal capture of the flow of affect. "Formed, qualified, situated perceptions and cognitions fulfilling functions of actual connection or blockage are the capture and closure of affect." (Massumi 1995, p. 96) It is with appraisal in the language of design that the transmission of affect is captured and felt.

Sentiment Analysis

Language Theory

In switching to the linguistic theory of appraisal, we will need to consider carefully how language is structured to display affect. Doing so will require us to consider the structures in language used by text to have and express emotions. Our assumption is that the language of design is reflecting the resonance of affect. Therefore, we need to account for the ways that language can display and have the resonance.

Understanding how language is used to construe affect has been theorized within the tradition of Halliday's theory of Systemic-Functional Linguistics (SFL) (Halliday 2004). According to Halliday, two linguistic features evoke appraisals: semantic meaning and grammar. One of the advantages of SFL theory is its emphasis on the syntactic structure of language to account for what language does and how the structure of language accomplishes various communication goals. SFL theory is concerned with the system of grammar within a genre of text and how the grammar produces meaning and relates experiences in the text. The text itself is considered to be strongly associated with a social situational context within which the text is produced. Within each specific genre, SFL theory holds that the system of grammar of a language constrains the choices available to generate meaning using language. SFL specifies a lexico-grammatical framework which constrains the features available to users of the language.

The theory of SFL and its associated analysis technique of functional grammar prescribe an analytical, criterion-based method for the functional-semantic analysis of the grammar and the participants in the grammar. The constraints imposed by the structure of a grammar yield the potential to analyze how the structural consequences of that choice relate to how language was used as a tool for representing knowledge or for making meaning (Halliday and Matthiessen 1999).

SFL theory specifies a set of grammatical systems used to produce meaning with language. Two of the systems relevant to our analysis are the TRANSITIVITY and APPRAISAL systems[2]. The system of TRANSITIVITY is the major system of grammatical choice involved in the way that people express experiential meaning, that is, in the way that people express reality. The system of TRANSITIVITY can be interpreted as the set of grammatical choices by which the language of design could be used to express the realities of the design process and the designed work. The TRANSITIVITY system is associated with the "ideational meta-function of language" (Halliday and Matthiessen 1999). To understand how grammar is implicated in the realization of the ideational meta-function of language, functional linguistics theory classifies clauses as processes in the TRANSITIVITY system. The processes contained within the transitivity system include: (i) material ('doing'), (ii) mental ('thinking'), (iii) behavioral ('behaving'), (v) relational ('being'), and (iv) existential ('existing'). I do not include the verbal ('saying') process as it is fairly limited in scope and can be subsumed within the other process types without much loss to how meaning was produced.

In contrast, the APPRAISAL system is associated with the meta-function of interpersonal exchange, as a medium of exchange between people. One could differentiate these two systems by thinking of the APPRAISAL system as negotiating attitudes about reality, which are expressed through the system of TRANSITIVITY. Appraisal theory in linguistics deals with the ways that speakers express evaluation, attitude and emotions through language. Linguists (Martin 2000; Martin and White 2005) define five high-level resources (Fig. 5.1) for conveying appraisals: attitude, engagement, graduation, orientation and polarity.

Attitude gives the type of appraisal. Within this category are three sub-types of attitudinal positioning: 1) affect – how the speaker is emotionally disposed to the

---

[2] Consistent with the notational standards of SFL, systems are designated in ALL CAPITAL LETTERS. Later, I will also use all capital letters to designate the PROCESS in functional grammar analysis in order to distinguish it from Process as a category of appraisal. However, PROCESS is not a system.

Fig. 5.1 Main linguistic resources
for appraisal and their attributes

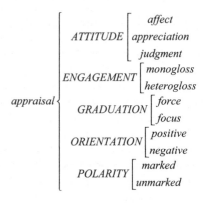

subject; 2) judgment – how the speaker assesses the subject with reference to behavioral norms and conventions; and 3) appreciation – how the speaker assesses
the form, aesthetics, and appearance of objects, including other people. Not sur-
prisingly, the semantic resource of Attitude is the resource most commonly associated with the expression of subjective content. The Attitudes in Table 5.1 are
typed according to Martin's framework.

Martin further classifies appraisals in terms of graduation and engagement.
Whereas attitudinal positioning describes the type of appraisal, graduation and engagement characterize the magnitude and the appraiser's commitment to the attitudinal position, respectively. Engagement is the commitment to appraisal and is
often considered an appraisal of the appraisal. It deals with subtle grading of the
speaker's commitment to what is said. Graduation deals with the strength of the
evaluation. Orientation relates to whether the appraisal is positive or negative. Polarity is labeled as marked or unmarked depending upon whether the appraisal is
scoped. The clause within which the semantic resource appears is an appraisal. That
is, any noun, verb, adjective, or adverb which functions to express meaning related
to affect, judgment or appreciation is considered a semantic resource of appraisal.

To illustrate these distinctions, let's take a look at some simple examples in Table 5.2 of appraisals and their use of the semantic resources.

If emotions are evaluative appraisals, then language is most revealing as a source
of difference in expressing emotions in design when we distinguish between
linguistic resources for appraising product, process (often called states of affairs
and happenings in linguistics), and people. It therefore becomes important to
know in the first instance to what external object the appraisal is directed toward.
Thus, it should be that the appraisals 'I am happy with the form of the product'

Table 5.1 Types of Attitude

Appraisal	Attitude Type
I am [happy] with the form of the product.	Affect
The form of the product is [pretty].	Appreciation
Joe is a [masterful] designer.	Judgment

Table 5.2 Sample appraisals

Appraisal	Semantic Resource and Attribute
This is a [good] design.	[good] (Attitude: appreciation; Orientation: positive)
This is a [great] design.	[great] (Attitude: appreciation; Orientation: positive; Graduation: force)
I [really] [think] this is a [great] design.	[really] (Graduation: force); [think] (Attitude: affect) [great] (Attitude: appreciation; Orientation: positive; Graduation: force)
This is a [good] design, [maybe].	[good] (Attitude: appreciation; Orientation: positive); [maybe] (Engagement: monogloss)
This is a [not] [good] design.	[not] (Polarity: marked); [good] (Attitude: appreciation; Orientation: negative)

and 'The form of the product is pretty' are the same subject category of appraisals, appraisal of a product, using the semantic resource of attitude, one with the attribute of affect and one with the attribute of appreciation.

Second, we need to account for the idea that emotions necessarily involve judgments about things which are of value to the designer whereas Martin's notion of judgment is limited to agents only. In design, judgments of objects and activities are routine; they are based on norms such as schools of design, theories of design, and accepted practices. Objects are judged in design and not just subjectively appreciated as Martin's framework prescribes. Thus, the appraisal 'The form of the product exemplifies Japanese aesthetics' is a judgment of the form of the product, not just a subjective appreciation. One could not make a legitimate appraisal for a product which does not imbue any Japanese aesthetics (unless you were being ironic) as agreed upon by a community.

We need to expand upon the notion of (human) affect in accord with Ortony's affective lexicon. That is, we need to deal with 'affect' as referring to affect, acting (behavior), or thinking (cognitive). We also need to deal with appraisals of 'designerly' capabilities. This category deals with appraisals of skill-based competencies and 'designerly' activities (Cross 1999), both of which are an integral aspect in appraising designers. That lets us then deal with judgment and appreciation as appraisals based on external or internal 'beliefs' rather than treating judgment as dealing with agents only and appreciation as dealing with objects and agents. The notion of judgment and appreciation is clarified: whereas external norms and standards are the basis for judgments, appreciation is based on interpersonal subjectivity. This clarification has a computational benefit in that knowing whether the appraisal is judgment or appreciation then differentiates the 'knowledge base' (as described in the OCC model of emotions) from which the appraisal is made: external (judgment) or internal (appreciation). It also affirms the object-relations differentiation between social, normative construction of judgment (external) and personal, background experiences as the basis for appreciation (internal).

Thus, I adapt Martin's appraisal framework to appraisal in the language of design in the following ways. At the top level are a set of semantic resources identifying the external stimuli producing the affect, Product, Process and People. I propose

Fig. 5.2 Categories of appraisal. The orientation of the appraisal is either positive (+) or negative (–). A neutral orientation is not considered an appraisal

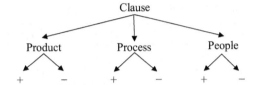

a hierarchical categorization in which clauses relate to situations that involve either the Product, Process, or People as shown in Fig. 5.2; that is, clauses in design text will engage in accounts about the Product of designing, Process of designing, or the People doing the designing.

Other researchers have differentiated these categories as process-oriented or content-oriented (Stempfle and Badke-Schaub 2002). Product-oriented thinking refers to reasoning about the goal space and the solution space of the designed artifact, about the function, form, behavior, and meaning of the artifact. Process-oriented thinking includes reasoning about activity and events. Activity refers to discipline-specific tasks and the collective process of teamwork (Valkenburg 1998). People-oriented appraisal may refer to cognition or meta-cognitive thinking and 'designerly' capabilities.

The categories of appraisal follow both the OCC cognitive model of emotions and the object-relations theorists' cognitive-evaluative view of emotions. The OCC model proposes that emotions arise as a consequence of valenced reactions to objects, events, and agents. Applying this idea to the framework depicted in Fig. 5.2, objects are the designed work (product), events are design processes (actions, activities and states of affairs that happen during design), and agents are people. The object-relations theorists view emotions as value judgments ascribed to objects and persons outside of a person's own control and which are of importance for the person's flourishing.

Second, the attributes of attitude are modified in definition according to the previous discussion: 1) affect – how the text uses affective, cognitive and cognitive-behavioral conditions to represent the stance toward the subject; 2) judgment – how the text appraises in relation to the accepted norms such as standards, industry best practices, established design methods, and objective criteria set by the design brief when the appraisal is of the Product or Process, or in relation to 'designerly' capabilities when the appraisal is of People; 3) appreciation – how the text appraises in relation to subjectivities defined by personal experiences.

A concern in this type of linguistic framework is whether the language actually reflects and/or indicates affective states and the extent to which the affective expression is sincere. Second, cognitive design researchers might question if the expression of a linguistic appraisal is concurrent with cognitive processing associated with affective appraisal. At minimum, we could argue that the intentional linguistic expression of appraisal is not possible without cognitively conscious affective appraisal processing in the brain on the assumption that the appraisal is sincere. It is possible that affective appraisals are occurring in the brain with no manifestation linguistically or otherwise. Any research which uses language, or

Fig. 5.3 Structure of the language of appraisal in design

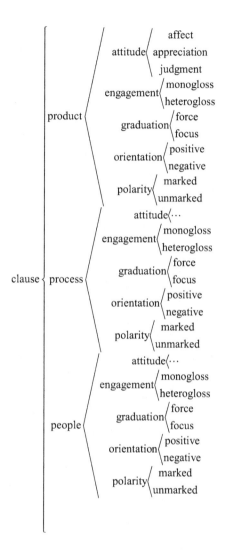

even certain physiological and bio-physiological indicators such as heart rate and skin conductivity, should acknowledge that we are only dealing with consciously available affective states. This framework can only answer how language construes emotion, not whether the emotion is 'truthful' or is cognitively grounded.

In summary, the structure of the language of appraisal (Fig. 5.3) prescribes a category classification and the semantic resources for conveying appraisals. The category classification is the aspect of design toward which the appraisal is directed, Product, Process or People. Appraisal is the representation through language of favorable and unfavorable opinions towards Product, Process or People. The semantic resources for appraisal and their attributes fall into each of the categories (Martin and White 2005). While the linguistic system of appraisal may not be isomorphic with human emotions, operationalizing emotions through a linguistic system allows us to begin an analysis of the affective aspects of the language of design.

Before I further explicate the details of the analysis of the language of appraisal in design, I would like to explain, briefly, the method of functional grammar analysis as the following sections utilize this analysis technique to illuminate the use of appraisals in the language of design. SFL's analysis technique of functional grammar analysis attempts to eliminate subjectivity in linguistic analysis by following a prescriptive, objective method for the functional-semantic analysis of the grammar and the participants in the grammar. While a full explanation of the SFL functional grammar analysis is beyond the scope of this book (see (Eggins 2004, pp. 206–253) for a detailed explanation), I provide a flavor for the analysis below in order to highlight the relatively high level of objectivity of the grammatical analysis. While necessary details about dealing with complex clauses, words that signify the PROCESS types and Participants, and handling appreciation versus judgment are dealt with later in this chapter, the example details the formal analysis techniques that will be undertaken to analyze the semantic-grammatical forms of appraisal in the language of design.

The examples will use the clauses 'It certainly takes an awfully long time to do stuff', 'It just looks unprofessional', and 'I don't regret changing them' to illustrate the analysis. The analysis proceeds as follows:

1. Identify the verb clause. This is known as the PROCESS.
2. Identify the Participants with the verb clause.

It	certainly takes	an awfully long time	to do stuff
Participant	Process	Participant	

It	just looks	painful
Participant	Process	Participant

I	don't regret	changing them
Participant	Process	Participant

3. Using the rules of the TRANSITIVITY system in SFL, decide the appropriate process type: mental (thinking), material (doing), relational (having, being), existential (existing) or behavioral (behaving) and the corresponding participant types. Code the Participant(s) according to their respective categories based on the PROCESS type. This analysis does not include the verbal PROCESS type as it is not significant in design text. For the purposes of this analysis, we consider relational and existential as equivalent. Below are the clauses from above analyzed using the TRANSITIVITY system.

It	certainly	takes	an awfully long time	to do stuff
Actor	Intensifier	Process: material	Range	Actor

It	just	looks	painful
Carrier	Intensifier	Process: relational	Attribute

I	don't	regret	changing them
Senser		Process: mental	Phenomenon: act

4. Identify the semantic resources for APPRAISAL. If there is no semantic resource evident, the clause is not an appraisal. Interpret whether the orientation of the appraisal is positive or negative. The semantic resources for APPRAISAL are indicated by [].

It	[certainly]	takes	an [awfully] [long]	time	to do stuff
Actor	Graduation: force	Process: material	Graduation: force; Attitude: appreciation	Range	Actor

It	[just]	looks	[painful]
Carrier	Graduation: force	Process: relational	Attribute, Attitude: appreciation

I	[don't]	[regret]	changing them
Senser	Polarity: marked	Process: mental; Attitude: affect	Phenomenon: act

5. Classify the clause as being about the design Product, the Process of designing, or the People doing the design. These three categories, Product, Process and People, describe the design situations (states) and events which may be the stimuli of affective judgment and therefore the subject of the linguistic appraisal.

We are now in a position to discuss methods to identify the semantic resources for appraisals (Step 4) and to categorize clauses (Step 5) using the PROCESS and Participant types.

Appraisal of Products

Appraisals of products are one way in which design text offer subjective assessments or apply normative judgments based on industry best practices, and external authoritative critique. Appraisals of products can justify (provide rationale) decisions taken during the design process. That is, their appraisals of products can explain how attitudes toward the designed object influenced the designing of the object. For example, the design text might express that a current design concept is better than a prior concept (Love 1999) or better than available products on the market. In the appraisal of the product, the text may rely on systems of linguistic resources that apply an external, normative judgment or appeal to subjective appreciation.

Whereas the appraisal of Process is associated with a tangible action, an appraisal of a Product is associated with an object. If this is based on formal analysis, we would consider this a judgment. If it is based solely on opinion, then it is a subjective appreciation. In the appraisals, the text can either accept or resist critique by others (where the others include design norms set by authoritative or legitimized sources) which is categorized as external valuation of products or provide a subjective appreciative stance on the designed object which is categorized as subjective valuation of products.

Further, the appraisals may examine distinct aspects of a product: form (the structure of the object), functionality (what the object does), behavior (how the object enacts the function), and the meaning of an object. Gero refers to the first three categories as the Function–Structure–Behavior schema for a designed object (Gero 1990). Valuations of meaning are what Rothery and Stenglin (1999) named "social value". These values may be extrinsically held or intrinsically held. If the value is intrinsically held, then the valuation is subjective regarding the meaning of the product to the design agent.

Appraisal of Product is not merely technical evaluation of the design work; rather appraisal is a means for evoking attitudes toward the work. Appraisals orchestrate expression of the characteristics of the design work by controlling whether and how vigorously various functional, structural and behavioral elements could become inscribed into the design work. Table 5.3 describes some appraisals of Product.

In the following examples, all linguistic labels are indicated by the use of SMALL-CAPS to distinguish with the design category of Process and where there might be ambiguity. Within clauses, appraisal clauses are indicated in *italic text*, where needed, with brackets [] indicating a semantic resource for appraisal.

The appraisals of the design work can consist of one or more pathways. One can appraise the design work by appraising the work directly as in the above examples in which there is generally a direct relation between the appraiser (X) and

Table 5.3 Appraisals of Product

Criterion	Example
The appraisal deals with the concept of the work	The visualization can be [appreciated] (Attitude: appreciation; Orientation: positive).
The appraisal deals with the structure of the work	I [really] (Graduation: force) [dislike] (Attitude: affect; Orientation: negative) any design that chops and changes between styles, layouts, formats, etc.
The appraisal deals with the behavior of the work	He said that there is [too much] (Graduation: force) user interactivity and that he [doesn't] (Polarity: marked Orientation: negative) [like] (Attitude: affect) the fact that the time disappears when the animation is over.

the appraised (Y) through a Relational-Attributive Process group or a Mental-Phenomenon Process group. However, it is also possible that the appraisal is expressed by describing a Mental PROCESS in one (separate) clause and the proposition or proposal in another clause. Take the following clause. Here, the speaker is describing her mental processing ("found"); however, it is the "it" (for which we are not given the anaphoric reference) which is being appraised.

I [never] (Graduation: force) [found] (Attitude: affect) it [intrusive] (Attitude: appreciation)

The clause could be rewritten as "I never found that it is intrusive." This is an example of a projecting clause which contains the mental PROCESS verb ('found') adjacent to another clause with the relational PROCESS verb (is). Such projecting clauses combine mental with relational clauses are often employed to appraise a Product or a Process rather than to describe a mental state. The appraisal clause is most commonly found adjacent to the clause "I think" or "I don't know". Take the following example:

[You know] (Engagement: heterogloss), [I mean] (Engagement: monogloss) if we can [perhaps] (Engagement: monogloss) have the visualization be [flexible] (Attitude: appreciation) so it can expand to accommodate [more] (Attitude: appreciation) visual elements so that the screen expands out like a sort of a cascade [or something] [Engagement: heterogloss] [I don't know] (Engagement: monogloss)

In this clause, the speaker is offering possibilities for the display area for a data visualization. She is hedging and appraising her suggestions by saying "I don't know" rather than making direct evaluative statements. In reporting on her knowledge state, she is simultaneously negotiating attitudes towards her suggestions and marking them as negotiable suggestions ('or something') rather than fixed and determinate design concepts.

Re-writing a passive Relational-Attributive into an active tense can be useful in identifying whether an appraisal is of Product or People. Take the following clause:

[You know] (Engagement: heterogloss) I'm [completely] (Graduation: force) [boggled] (Attitude: affect) by the blinking squares in the visualization

In the active sense, this clause would have been "The blinking squares in the visualization completely boggle me." Here, it is the "blinking squares" which are doing the action of boggling; since boggling is considered a negative action, it is the blinking squares which are being appraised, rather than the mental state of "me".

The evaluation of the design work is generally integrated into a description or technical analysis of the design work. However, we only code an evaluation as appraisal when the language is expressing the speaker's stance toward the description and evaluation. In the following clause, the speaker is describing the visualization and what it can do. The clause can also be interpreted as evaluating the visualization in terms of its structure. It is only when he mentions that you get a "lovely feel" from the visualization that we get an indication of his stance toward the data visualization.

> The visualization if you like has bird objects or boids represented as circles and they stand for each person in the team and the speed with which the boids flock together give you an indication of how the team is working together and over time you get a [lovely] (Attitude: appreciation) feeling watching the [rhythmic] (Attitude: appreciation) motions of the boids as they careen toward and away from each other

One of the more problematic aspects in coding appraisals of product is dealing with projection of possibilities (Dong et al. 2005). When a participant is offering a potential solution or needed characteristic of the design work, does it always signal the person's attitude toward the proposal? Does offering a suggestion not indicate someone's preferred concept? We do not categorize a projection of a possibility as necessarily the expression of an attitude towards the possible solution unless the expression includes an explicit linguistic signal ascribing the speaker's attitude toward the possibility. Take the following clauses:

> I [think] (Attitude: affect) you [absolutely] (Graduation: force) need to use [higher] (Graduation: force) contrast colors

This clause is a projection of possibility in which the designer suggests the need for "higher" contrasting colors, signaling his appraisal that the current color choice is lacking in contrast. Compare this clause with the following projection of possibility with no attitudinal positioning:

> I think linking it to the database or something

Here, the designer is offering one solution to connect the visualization to the data source via a relational database, but he does not indicate his preference or distaste for this possibility. As such, it is not considered an appraisal.

Appraisal of Process

It is generally regarded in design management that there exists four key aspects of successful design (or product development) in industry: 1) Strategy; 2) Organization; 3) Processes; and 4) Tools. Regardless, the appraisal of strategy, organization, processes and tools valuates either their capability to support, facilitate, or perform 'designing' or their current state of affairs. The appraisal of PROCESS is identified by taking stances towards tangible tasks and actions. If the appraisal makes references to established norms in design practice, then the appraisal is considered as judgment. Conversely, if the appraisal relies on experience and interpretations of the situation, it is considered a subjective appreciation.

In all of the PROCESS-oriented appraisal clauses, a tangible action is being evaluated, not the agent undertaking the action. The evaluation associates a position toward the state of being of the action. Appraisals of PROCESS can normally be identified by asking, 'What is being/was done?' and then 'What is/was the stance toward the action?' How does the design agent feel about a potential outcome of a design move (Langan-Fox and Shirley 2003)? The following questions in Table 5.4 guide the identification of appraisals towards design activities.

Table 5.4 Appraisals of Process

Criterion	Example
The appraisal is taken toward a specific task or action	This option puts me [ill-at-ease] (Attitude: affect) if I pursue it.
The appraisal is commenting on the need for an action	We'll [have] (Graduation: force) to describe the visualization algorithm [in detail] (Graduation: force) in the white paper.
The appraisal is taken towards generic design processes	This is [not] (Polarity: marked) the design process at its [finest] (Attitude: judgment; Graduation: force).

One problem commonly encountered in coding appraisals of Process occurs when the appraisal of Process stands in for an appraisal of Product. In the following utterance, the designer apparently appraises a task.

We've just very simply clustered the data objects in the visualization by color.

If we were to follow the criterion above, this would be categorized as an appraisal of Process. Yet, we might ask ourselves, is the speaker really intending to appraise the simplicity of the action of grouping the data objects or is it that the data objects have been grouped simply, and thus their simple grouping is an attribute of the data objects in the visualization. To make the distinction, we can rewrite the clauses by inverting the Participants and re-writing the clause into a Passive form such as "The data objects have just been very simply grouped by us." This inversion shows us more clearly that the simple grouping by the actors ("us") is an attribute of the data objects. While appraisals of Process can be identified by asking, 'How did X do Y to Z?' as in "We (X)'ve just very simply grouped (Y) the data objects (Z)", in these cases, there is a closer relationship between the PROCESS ("grouped") and the second participant ("the data objects"). In Halliday's terminology, he called the second Participant Ranges. A Range specifies the domain or scope of the PROCESS, is often semantically closer to the PROCESS, and specifies a part of the PROCESS. That is, a Range either restates or continues the PROCESS itself or the range of the PROCESS. In this example, it is not possible to group without scoping the domain of grouping; so, the data objects are just the domain of the PROCESS of the material process of grouping (Eggins 2004, p. 218). Where this closer relationship exists and the second Participant refers to the Product, we code these as appraisals of Product rather than Process; in these cases, the second Participant is a Range rather than a Goal.

Appraisal of People

Appraisals of People express subjective valuations of a person's (a stakeholder in the design process) cognitive and physical states of being and capabilities. Appraisal of states of being includes internal, mental states of affect, cognitive states, cognitive-behavioral states, and physical functioning or capability (the functioning of or capability to undertake a tangible design-oriented action).

Table 5.5 Appraisals of People

Criterion	Example
The appraisal judges social esteem (normality)	I'm [just] (Graduation: focus) [very] (Graduation: force) [fussy] (Attitude: affect)
The appraisal judges social esteem (tenacity)	I've been trying my [hardest] (Attitude: judgment) I was [so] (Graduation: force) [in the zone] (Attitude: appreciation).
The appraisal judges social sanction (ethics)	We as architects have [not] (Polarity: marked) been [that] (Graduation: force) [forward thinking] (Attitude: judgment) [in times like these] (Engagement: force) towards environmental sustainability.

Appraisals of People are generally associated with the Mental and Behavioral PROCESS or the Relational PROCESS where the Carrier is a sentient being and the second participant is Attributive or Identifying. One of the major challenges is coding what counts as appraisals of People, since all first-person and third-person descriptions of People could be construed as advancing an opinion. Descriptions of People tend to take on an air of normative evaluation about how people should and should not be or behave. Following the standard practice in SFL analysis, I also applied Iedema's criteria (1994) of social esteem (normality, capacity or tenacity) and social sanction (truth and ethics) to identify appraisals as judgments of People. These criteria can also be cross-referenced with Nigel Cross's concepts of 'designerly' behaviors and 'natural intelligence'.

In building up an outline of the type of appraisals of a design agent, consider the factors in Table 5.5.

Previously, I had discussed projecting clauses. A projecting clause that places a Mental PROCESS adjacent to a Material or a Relational PROCESS is likely to be an appraisal of Product or Process rather than People. Describing a mental state directed towards the subject of appraisal is an appraisal of the subject, not of the designer. Take the following examples.

> I (Carrier)'{m} (Process: relational) [very] (Graduation: force) [interested] (Attitude: affect) to [sort of] (Graduation: force) {keep} (Process: material) the sound dampening (Range) in the design concept

Being "keen" is the mental state, but it is directed toward "keep[ing] the sound dampening" rather than appraising the designer.

> I (Senser) [{wonder}] (Process: mental; Attitude: affect) if it (Carrier)'{s} (Process: relational) possible (Attribute: intensive) to {manufacture} (Process: material) something like that (Goal)

To "wonder" is the Mental PROCESS, but it is directed (through the projected clause) toward appraising the possibility of manufacturing the design artifact. This clause is different from the following, which is an appraisal of mental state because the second participant is the Phenomenon of the Mental PROCESS.

> They (Senser)'{re} [just] (Graduation: focus) [{wondering}] (Process: mental; Attitude: affect) how the finish of the metal will look (Phenomenon)

It is usually possible to inquire whether a clause, which contains a description of a mental state, is an appraisal of Product or Process, if it is possible to ask the opinion of the Carrier of the Phenomenon or Circumstance relating to the first clause in the projected clause.

The directors (Carrier) will {be} (Process: relational) [slightly] (Graduation: force) [interested] (Attitude: affect) in this concept (Circumstance: matter) but they (Carrier) are (Process: relational) [only] (Graduation: focus) [interested] (Attitude: affect) in how much they'll cost (Circumstance: cause)

The reverse situation is also possible, where a Material PROCESS is projected onto a Relational PROCESS clause to appraise.

So we (Actor) should {try} (Process: material) and {be} (Process: relational) a [bit more] (Graduation: force) [disciplined] (Attitude: judgment) with the material choices

The connotation in this appraisal is that the group is not disciplined.

We are now in a position to characterize the linguistic resources used to express appraisals and the subject toward which the appraisal is directed, Product, Process or People. The following rules can be used to ascertain the top-level category of the appraisal from the functional-grammar parse.

1. As only a conscious agent or an anthropomorphized non-human may have a Mental or Behavioral PROCESS type, it is required to differentiate between the grammatical forms of statements such as 'I like this design' (Process: mental; Attitude: appreciation; Category: Product) and 'I feel disappointed' (Process: mental; Attitude: affect; Category: People). As such, the second participant, called the non-active Participant or Phenomenon, is used to distinguish the category. Note that the appearance of a non-cognitive agent as the first Participant in a Mental PROCESS is always invertible into a passive clause. So, 'The visualization (Senser) [completely] (Graduation: force) [{frustrates}] (Process: mental; Attitude: affect) me (Phenomenon)' is equivalent to 'I'm [completely] (Graduation: force) [frustrated] (Attitude: affect) by the visualization'.

   1. If the Senser is an anthropomorphized non-human, and the Phenomenon is a design stakeholder, then the clause is a Product clause. This construct is not present in this data set, but one could quite easily imagine the following clause: Computers (Senser) do [not] (Polarity: scoped) [{like}] (Process: mental; Attitude: affect) me (Phenomenon). This colloquial saying is generally intended to signal a disdain for computers rather than to appraise the person, "me", as unlikeable.
   2. If the Phenomenon or a projected clause for which the first Participant is a circumstance about the design work according to the criterion for appraisals of Product, then the category is Product. Example: I (Senser) [quite] (Graduation: force) [{like}] (Process: mental; Attitude: affect) the idea of a [smaller] (Graduation: force) motor (Phenomenon)
   3. If the Phenomenon or a projected clause in which the first Participant is a circumstance about the design process according to the criterion for ap-

praisals of Process, then the category is Process. Example: I (Senser) [{think}] (Process: mental; Attitude: affect) we (Actor) [have] (Graduation: force) to {do} (Process: material) a [little bit more] (Graduation: force) technical research (Goal)

4. If the Phenomenon is an affect, capability, or a design stakeholder, then the category is People. Example: I (Senser) [{like}] (Process: mental; Attitude: affect) Chris (Phenomenon)

2. If the PROCESS is Material, then we must examine the semantic category of the Actor and the Goal or Range simultaneously. Specifically, we have to pay attention to whether the second Participant is closer to the PROCESS than the first Participant. If the second Participant or the projected clause is closer to the PROCESS, that is, it is a Range, then the category is determined by the second Participant.

   1. If the Actor is circumstance about the design work, the clause belongs in the category Product. The housing (Actor) {can make} (Process: material) it (Goal) look [sturdy] (Attitude: appreciation) but in an [inexpensive] (Attitude: judgment) way.

   2. If the Actor is a design stakeholder and the Range is a circumstance about the design work, the clause belongs in the category Product. We (Actor)'re going to [have] (Graduation: force) to {reduce} (Process: material) the cost (Range) [substantially] (Graduation: force)

   3. If the Actor is a circumstance about the design process, the clause belongs in the category Process. Acoustic dampening (Actor) will {help} (Process: material) reduce vibration [enormously] (Graduation: force)

   4. If the Actor is a design stakeholder and Goal or Range is a circumstance about the design process or the result of the Goal refers to the manner in which the Goal was achieved, the clause belongs in the category Process. You (Actor)'d [have] (Graduation: force) to {remove} (Process: material) the housing (Goal) to [actually] {reveal} (Process: material) the motor (Range) [in the first place] (Graduation: focus)

   5. If the Actor is a design stakeholder and the result of the Goal or the manner of achieving the Goal (i.e., the manner of the Material PROCESS) refers to the Actor, then the clause belongs in the category People. We (Actor)'ve [only] (Graduation: force) {modified} this (Goal) [once] (Graduation: force) [over the product life-cycle] (Graduation: force)

3. If the PROCESS is Relational or Existential, then we must examine the semantic category of the Carrier. The second Participant is either Attributive, referring to a quality of the Carrier, or Identifying, defining the Carrier. Because Relational or Existential clauses can project, we must also pay attention to the projected clause and apply the same criterion.

   1. If the Carrier or projected clause is a circumstance relating to the design work, then the clause belongs in the category Product. It (Carrier) will {be} (Process: relational) [very] (Graduation: force) [quiet] (Attitude: judgment) and [energy efficient] (Attitude: judgment) sounding (Attributive)

2. If the Carrier or projected clause is an act, event, phenomenon, or state, then the clause belongs in the category Process. I (Carrier)'{m} (Process: relational) [very] (Graduation: force) [insistent] (Attributive; Attitude: affect) to [sort of] (Engagement: monogloss) {keep} (Process: material) that in there

3. If the relation between the Carrier and the second Participant is affective, cognitive or behavioral, and the clause is followed by a subordinate clause in the form of a prepositional phrase, then we must check what the subordinate clause refers to. The following quote from Natalia Ilyin illustrates this pattern: "I (Carrier)'{m} (Process: relational) tired (Attitude: affect) of the [narrow] (Graduation: force) language, the [small] (Graduation: force) sandbox, the [limits] (Attitude: appreciation) of what we deem 'good design.'" Ilyin is not appraising herself feeling tired; she is appraising what is considered design (Category: Product).

4. If the Carrier or projected clause is a person, body, feeling, or cognition, then the clause belongs in the category People. I (Carrier)'m (Process: relational) a [little] (Graduation: force) [inexperienced] (Attitude: judgment) (Attributive)

The linguistic analysis of appraisal is usually quite complicated due to the complex clauses found in the text and the semantic interpretation of metaphors of appraisal rather than straightforward adjective and adverbial modifiers of nouns and verbs to express sentiment and polarity. The following is one such complex example.

> I'm going through my design and doing some [last minute] polishing, and [desperately] trying to make my documentation [slightly less nonsensical] than what I thought was a [literary masterpiece] at 3am.

This single sentence contains an appraisal of Process and Product, but the positive appraisal of the Product is based on a metaphor – and one might question whether the student would still appraise the documentation as a "literary masterpiece."

Other challenges arise in coding the appraisals which merit some discussion. The following examples are taken from the Ivan, John and Kerry tape from the Delft Protocol Workshops. The functional grammatical analysis technique is complicated by many possible textual realizations for registering appraisals beyond a single clause or through the use of collective beliefs of what is 'good' (such as more is better than less, faster is better than slower, etc.), and specific grammatical forms such as comparison. All of these are, of course, culturally and contextually bound; knowing the orientation of the appraisal will require a bit of 'local knowledge'. Here are linguistic examples of appraisal that we might find in design text which do not readily follow the above rules.

1. Appraisal through repetition – In segment (t 1503), John appraises his idea of "little snaps" by quantifying its merits through multiple (three) justifications flagged by the conjunction term "so", one justification flagged by "then", and finally three justifications flagged as actions "you" could do given his proposed solution. The number of reasons John delineates clearly expresses his positive orientation toward the snaps.

and this could [just] have little snaps to the er er to these rails [so] that to these tubes [so] we
have this folding down spec [so] that if this junc point here had a pivot at it [and and then] it's
kinda like you're folding TV trays you [just] unclip this guy from here [and] you unclip well
you probably don't need to unclip the back one you [just] unclip one of these [and then] you
can swing the legs flat

2. Appraisal by reference to time scales – Time is critical in product development
   and is thus a marker for Attitude when it signals urgency. John uses this linguistic resource to express his negative orientation toward not having identified
   the needed parts in (t 1668):

   so with that we're like two parts plus straps [right now] but we don't we don't have this crit-
   ter identified [now]

3. Appraisal by comparison – By comparing aluminum to steel tubing in (t 1066),
   John signals his attitude against steel tubing.

   [if] we used um aluminum tubing [instead] of er instead of steel tubing [not only] is there
   a weight savings but we could er meet the ugly spec

Sentiments are thus evoked by two carriers: semantic meaning and grammatical
features. Affective specifiers are closely linked to subjective words. Adjectives
and adverbs which are positive or negative modifiers can be identified according
to the General Inquirer corpus [http://www.wjh.harvard.edu/~inquirer/]. However,
subjective content can be carried through nouns and verbs as well. The problem in
identifying semantic content that carries affective content is that reading subjective content from design text creates a coding problem. Formally identifying all
possible ways affect could be keyed linguistically is a daunting if not altogether
impossible task, likely to miss ways of expressing affect. To assist in this process,
it is useful to create a table identifying the common semantic and grammatical
structures used by language to appraise.

Following APPRAISAL theory, a group of words to assist in identifying affective
content is summarized in Table 5.6. The sample words for the semantic resource
of affect, an attribute of attitude, were taken from the OCC affective lexicon
(1987). While the OCC affective lexicon has five categories – affective, affective-
behavioral, affective-cognitive, behavioral-cognitive, and cognitive – we reduced
them to three since affect is either the sole or predominant focus of two categories.
Thus, affective, affective-behavioral, affective-cognitive are a part of the affective
category, behavioral-cognitive is behavioral, and cognitive is cognitive. All three
are attributes of the semantic resource of Attitude. To identify semantic realizations of affect, I follow the prescription set by Ortony's affective lexicon. The
affective lexicon distinguishes these three categories based on the "significant
referential focus" of a word. Affective words express internal, mental states of
being which do not have a significant cognitive or behavioral focus. Words with
a cognitive focus "refer to aspects of knowing, believing or thinking. Specifically,
they refer to such things as readiness, success, and desire." Words with a cognitive-behavioral focus refer to how a person is "thinking about a situation as well as
to how one is acting." (Ortony et al. 1987, p. 352)

At least two layers of meaning exist in clauses, a layer of meaning associated with construing reality, through the system of TRANSITIVITY, and a layer

Table 5.6 Sample words keying appraisals

Semantic Resource	Functional Grammar Type	Sample Words
Affect	Process: Mental	like, dislike, enjoy,
	Process: Relational	ill-at-ease, dissatisfied, apprehensive
Behavioral	Process: Mental	hesitate, care, defy
	Process: Relational	careful, cautious, lazy, stupid, silly
Cognitive	Process: Mental	accept, tolerate, expect
	Process: Relational	aware, amazed, certain, confident
Graduation	Graduation: Force	very, really, extremely, at the moment, recent, right now
	Graduation: Focus	in particular, effectively,
Engagement	Engagement: monogloss	in my opinion, in my view, I believe
	Engagement: heterogloss	it is said, so to speak, it seems, probably, perhaps, maybe, sort of

associated with construing emotions, through the system of APPRAISAL. The TRANSITIVITY system and a modified version of the APPRAISAL system from the theory of SFL provide a rigorous manner by which appraisals can be identified. The examples have shown linguistic appraisals, publicly available emotions, in the language of design. The analysis includes identifying the semantic and grammatical structures of emotion and how these structures express affective content. In coding linguistic appraisals in design text, we are not merely looking for evaluative statements. Rather, our interest is in how the language of appraisal in design is functioning to negotiate attitudes toward the design work, the design process, and the people doing the design work. Specific grammatical forms are associated with appraisals in the language of design. The rigor opens the potential for the implementation of computational appraisal parsers that might be useful in detecting the instantaneous changes of emotions in the language of design.

Computational Implementation

While the method of functional grammatical analysis is applicable for a close examination of how the grammar 'works' to generate attitudes, it is often not practicable for the large-scale analysis of language. Given that robust and accurate computational functional grammatical analysis of language is currently unavailable, we need to find an alternative means to analyze attitudes in text. However, the functional grammatical analysis offers us some guidance about how we might go about building an attitude pattern detection machine.

Whereas the computational linguistic methods of latent semantic analysis and lexical chain analysis are deployed on problems of semantic meaning, understanding attitudinal positions in text is a separate problem altogether. To illustrate why semantic meaning is not sufficient to register attitudes, let's consider the following exchange between the participants of the Delft backpack design team:

> J (t 557) like where they want that mass maybe the if there's a thing that comes down to here you could have it so that it adjusts so you could kinda lever the pack up or down a little bit y'know if it's not a a fixed
> K (t 558) seems like lower is better regardless as you say like we design in the low position and not necessarily try and get
> J (t 627) if it was a smaller article it would work but not if it's something this size um …
> and over the front does do people have any problems with mounting it up front?
> I (t 628) yeah you have more mass up there to turn

If we were only to deal with the semantic meaning of this exchange, we could extract that they are discussing the placement of the backpack, which is referred to as "mass" and "article" which are both semantically related by the concept 'physical object'. However, their attitude toward the placement of the backpack is summarized by the appraisal "lower is better regardless". Without the appraisal, an understanding of the preferred solution would not be available.

Understanding appraisal is another layer in understanding the intention of a sentence and is separate from semantic meaning. The term semantic orientation is used to differentiate between these two layers of understanding. In computational linguistics, the term sentiment analysis is used to describe the class of algorithms that register the semantic orientation of text. Sentiment analysis tries not only to register what the attitudinal position was, that is, either positive or negative, but also the 'size' of the appraisal and the appraiser's commitment to the appraisal. The computational implementation described in this section focuses on two resources in the language of appraisal of design, the category of the appraisal and the orientation of the appraisal, and their attributes, Product, Process or People, and positive or negative, respectively.

One of the major problems with sentiment analysis is the identification of the semantics and grammatical structures of appraisals. In the examples provided thus far, the appraisals were expressed as either adjective–noun pairs or single terms which carry an attitudinal position (such as 'good' as in 'This is good'). If appraisals were expressed in such a simple unigram or bigram structure, then the problem becomes one of identifying the orientation of the terms which register the attitudinal position. For example, we would want to know that the term 'good' is a term of positive attitude and 'bad' is an expression of negative attitude.

Turney (Turney 2001; Turney and Littman 2003) developed two techniques to infer the semantic orientation of a term based on the term's statistical co-occurrence with terms known to express positive or negative orientation. That is, if the term 'fabulous' occurs more often with the known positive term 'good' than with the known negative term 'bad', then we would infer that 'fabulous' is also a term of positive orientation. They computed the semantic orientation of a term as the difference between the strength of a term's association with a set of positive words minus the strength of its association with a set of negative words. The statistical measure used to measure the strength of a term's association with the set of positive or negative words is called Pointwise Mutual Information (PMI) (Eq. 5.1). The Pointwise Mutual Information – Information Retrieval (PMI–IR) calculates the log-odds of the co-occurrence of the target word ($word_1$ in Eq. 5.1) with a known positive/negative word ($word_2$ in Eq. 5.1; $pword$ or $nword$, respectively, in Eq. 5.2) in an information retrieval (IR) corpus. The value of the PMI–IR is calculated by issuing a query to a search engine to count the number of hits (matching documents) which contain both words. Then, the value of PMI is calculated using Eq. 5.1.

$$PMI\left(word_1, word_2\right) = \log_2\left(\frac{p\left(word_1 \,\&\, word_2\right)}{p\left(word_1\right)p\left(word_2\right)}\right) \quad\quad (5.1)$$

Turney and Littman derive the Semantic Orientation–Pointwise Mutual Information (SO–PMI) metric based on the number of documents returned by a query to the search engine (hits) as:

$$SO - PMI\left(word\right)$$
$$= \log_2\left(\frac{\Pi_{pword \in Pwords}\, hits\left(wordNEARpword\right) \cdot \Pi_{nword \in Nwords}\, hits\left(nword\right)}{\Pi_{nword \in Nwords}\, hits\left(pword\right) \cdot \Pi_{nword \in Nwords}\, hits\left(wordNEARnword\right)}\right) \quad (5.2)$$

With the Google search engine, the NEAR operator can be handled with the * operator. That is, the search `design * good` finds all documents in which the word design is separated by the word good by one or more words. Since the numerator is a constant, it only needs to be calculated once for a given canonical set of positive words ($pword$) and negative words ($nword$).

In practice, one can locate a canonical set of canonical positive and negative words based on large-scale word usage catalogs. One such catalog is the British National Corpus (Leech et al. 2001, pp. 286–293) which contains the frequency of occurrence of words in written and spoken English. To ascertain which of these words is positive or negative in orientation requires cross-referencing them with another corpus. The General Inquirer corpus tags modifiers, adjectives and adverbs, as positive or negative modifiers. Crossing these two lists together, it is possible to derive a canonical set of frequently occurring words with known positive or negative orientation. This is useful for calculating the SO–PMI. The basis for the selection of these frequently occurring words as the canonical set is the increased likelihood of finding documents which contain both the canonical word and the word for which the SO–PMI is being calculated. This increases the accuracy of the SO–PMI measurement. Table 5.7 lists the canonical adjectives their frequency per million words[3].

---

[3] The adverbs *just* and *too* appear 1277 and 701 times per million words and are negative according to General Inquirer. That would put them both on the list of positive and negative words. However, we do not include these words due to their use to graduate appraisals such as 'too good' or 'just dangerous' rather than being able to express an opinion on their own. As such, we do not include any adverbs which apply graduation to appraisals.

Table 5.7 Canonical positive and negative words

Positive Words	Negative Words
good (1276)	bad (264)
well (1119)	difficult (220)
great (635)	dark (104)
important (392)	cold (103)
able (304)	cheap (68)
clear (239)	dangerous (58)

In practice, the SO–PMI values for words are calculated in a batch process and stored in a hash table. When the machine learning algorithm goes to look up the SO–PMI value, if it does not exist in the hash, then the value is calculated and added to the hash table.

In Turney's empirical studies, he showed SO–PMI can achieve accuracy better than 80% in classifying the correct semantic orientation of a text. However, one of the main shortcomings of Turney's co-occurrence based technique is its dependency on a set of terms which are context-free with respect to semantic orientation. Second, having to query a large corpus to measure the strength of association is computationally expensive. The cost of putting together such a corpus should not be understated either[4]. While much of the lookup and calculation could be done off-line, this requires maintaining a large database of terms and their statistically derived semantic orientation.

An alternative approach has been investigated by Lillian Lee (Pang et al. 2002) and her students at Cornell University. They compared a set of supervised machine learning algorithms, Naïve Bayes, maximum entropy classification, and support vector machines, in their ability to successfully classify positive and negative movie reviews. Their study aimed to ascertain whether these supervised machine learning methods, which have been applied successfully to text categorization, could be equally successful at categorizing semantic orientation, which is similar to text categorization if positive and negative orientation is considered like a 'thematic' category of text. They found that the three supervised machine learning algorithms performed about as well as Turney and Littman's system, with support vector machines performing better than both Naïve Bayes and maximum entropy classification. What is most intriguing from Pang's study is that the best performance resulted from a sentiment classifier that relied only on a bag-of-words representation of the design text and support vector machines.

The best performing sentiment classifier (i.e., the most accurate) is based on characterizing a document as a bag-of-words with no other knowledge of the semantic category or the semantic orientation of the words included. The bag-of-words representation of the text using unigrams found in the text (Pang et al.

---

[4] Fortunately, Google has provided their n-gram database to the Linguistic Data Consortium for distribution. The Linguistic Data Consortium provides a database of n-grams called the Web 1T 5-gram. This set of n-grams could be used to calculate the SO–PMI.

2002) does not include any knowledge about feature sets from the semantic resources in the APPRAISAL system. Each document in the corpus is numerically represented as a high-dimension vector of frequency counts of unigrams. For any unigram in the candidate set of content-bearing bag-of-words, if the unigram appears in both the document and the list, the corresponding position in the vector will be labeled as 1; else it will be marked as 0. That is, the representation of the text is exactly the same as the word-by-document representation used for latent semantic analysis. This bag-of-words sentiment analysis method attained almost 90% accuracy in semantic orientation classification. Only by combining the semantic resources for Attitude and Orientation from Martin's APPRAISAL system into 'appraisal groups' with bag-of-words features does the performance of sentiment classification improve (Whitelaw et al. 2005), though not by much. Another method which employed cognitive and commonsense knowledge as rules also does not perform as well as the bag-of-words classification (Shaikh et al. 2007).

Whilst it is not altogether explainable, at least theoretically, why the bag-of-words method works so well compared to methods rigorously grounded in linguistic theory, it is tempting to apply the bag-of-words classifier immediately to design text nonetheless. After all, the performance of Whitelaw's APPRAISAL theory based classifier is not so markedly better to justify the additional computational expense of generating the lexicon for the appraisal groups. However, it is important to note that the bag-of-words classifier was trained on a data set consisting of text about movie reviews. It is not *a priori* obvious whether a supervised machine learning sentiment classifier system trained on a data set from a specific domain, movie reviews, could perform equally well on another domain, design text. The lexicon of the two domains is different. It is not known if this difference will result in a significant degradation in performance in accuracy of sentiment classification when the classifier is trained on one data set but deployed on another. This drop in performance has been claimed (Shaikh et al. 2007) but not tested. There is the potential that it might not be necessary to utilize all the words that appear in the target domain to train the classifier. Given the evidence from research by Wiebe (1999) that a sentence is 55.8% likely to have subjective content if there is an adjective within, the appearance of a broad range of adjectives in the training data set might be sufficient for sentiment analysis. It would be highly attractive to reuse a sentiment classifier trained on existing data sets due to the expense of producing tagged data to train machine learning classifiers. However, it is not likely that it would be plausible to train a sentiment classifier on one corpus, such as the movie review data set used by Pang, and then deploy the classifier on another corpus, such as the design text associated with a specific project. Nonetheless, the simple representation of text of the bag-of-words sentiment classifier, which is consistent with the representation for latent semantic analysis, makes it an attractive method for sentiment analysis.

One key finding from current research in sentiment analysis is that improvements in correctly identifying the correct attitudinal position will not lay in better identifying the semantic orientation of words in a text but rather by first ascertaining whether the attitudinal position is related to the topic of appraisal. That is, the

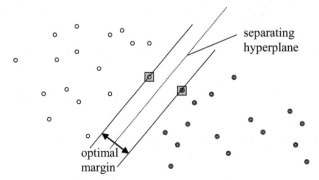

Fig. 5.4 Separating hyperplane in a two-dimensional space. The *gray boxes* indicate the support vectors. The support vectors define the margin of largest separation between the two classes

attitudinal position must be referenced to the appropriate topic and merely examining the semantic orientation of a 'bag-of-words' in a text is not sufficient to correctly identify the semantic orientation of the text. Given the demonstrated success of support vector machines for both text categorization (Joachims 2002) and semantic orientation classification (Pang et al. 2002), we will use them as the supervised machine learning algorithm to train a system to identify both the category and orientation of design text.

A support vector machine (Cortes and Vapnik 1995) is a supervised machine learning technique that finds the basis for the optimal separation of high dimensional vectors into two classes. Used typically for classification problems, input support vectors are mapped into a high dimensional feature space to find a hyperplane that separates the vectors by the largest margin. The support vectors are the training data which are used to calculate the margin. The optimal hyperplane is defined as the linear decision function with maximum margin between the vectors comprising the two classes. Graphically, the hyperplane separating the data into two classes is illustrated in Fig. 5.4. We can imagine that all documents with positive orientation are on the left-hand size of the separating hyperplane, negative orientation on the right.

Mathematically, finding the hyperplane is a constrained optimization problem. Let $\mathbf{w}_0 \mathbf{x} + b_0 = 0$ define the optimal hyperplane. Let $\mathbf{x}$ be the support vectors and $\mathbf{y}$ be the labels for the support vectors, that is, $\mathbf{y}$ defines the class to which a support vector belongs. For a set of patterns $\mathbf{x}_i \in \mathfrak{R}^n$ with labels $y_i \in \{-1, 1\}$ that are linearly separable in the input space, the separating hyperplane must satisfy for the set of labeled training patterns $(y_i, \mathbf{x})$ (i.e., the training data) a set of constraints expressed by the equations:

$$\mathbf{w} \cdot \mathbf{x}_i + b \geq 1 \text{ if } y_i = 1,$$

and

$$\mathbf{w} \cdot \mathbf{x}_i + b < -1 \text{ if } y_i = -1.$$

These constraints are normally written in the form:

$$y_i \, (\mathbf{w} \cdot \mathbf{x}_i + b) \geq 1, \; i = 1 \ldots l$$

where l is the number of labels (size of the training set). The inequality constraints become equality constraints for points from the two classes that are closest to the hyperplane. The vectors $\mathbf{x}_i$ where the inequality constraint becomes active, i.e., $\mathbf{w} \cdot \mathbf{x}_i + b = 1$ are called the support vectors. Conveniently, the labels $y_i \in \{-1, 1\}$ correspond to a positive (appraisal) orientation when $y_i = 1$ and a negative orientation when $y_i = -1$.

The distance between the projections of the training vectors of the two different classes is given by the equation $2/\|\mathbf{w}\|$. Cortes and Vapnik prove that the optimal hyperplane is the one which maximizes the distance between the points on either side of the hyperplane and for which the inequality constraints become equality constraints. Maximizing the distance is equivalent to minimizing the inverse of the distance, given by the function $\frac{1}{2}\mathbf{w} \cdot \mathbf{w}$, subject to the above constraints. Thus, the problem can be recast as a constrained optimization problem. We can re-write the optimization problem in the standard Lagrangian form as:

$$L(\mathbf{w}, b, \lambda) = \tfrac{1}{2}\mathbf{w} \cdot \mathbf{w} - \sum_{i=1}^{l} \alpha_i \left[y_i (\mathbf{x}_i \cdot \mathbf{w} + b) - 1 \right]$$

Finding the point on the Lagrangian where the slope goes to zero by taking the partial differentials with respect to $\mathbf{w}$ and b and solving the resulting equations at the saddle point $(\mathbf{w}_0, b_0)$ we derive the following equation.

$$\mathbf{w}_0 = \sum_{i=1}^{l} \alpha_i y_i \mathbf{x}_i \tag{5.3}$$

The equation expresses that the optimal hyperplane solution can be written as a linear combination of the support vectors.

Sample Study

We thus approach sentiment analysis of design text as a dual problem of semantic meaning and semantic orientation. The sentiment classification proceeds in two stages:

1. Determine the category of the text, i.e., Product, Process or People ($\mathrm{SVM}_{category}$)
2. Within each category, determine the semantic orientation toward the category ($\mathrm{SVM}_{sentiment}$)

In order to make use of the support vector machines, we need to represent the design text in a vector format $\mathbf{x}_i$. Again, we will use the same word by document representation as for latent semantic analysis. We will call this sentiment classifier a 'bag-of-words sentiment classifier' since it represents the target clauses as,

Fig. 5.5 Sample vector representation of the
text data $\mathbf{x}_i$ and the class label (orientation) y_i
used in a support vector machine

$$\left[\underbrace{+1}_{y_i}\ \underbrace{0\ \ 0\ \ 0\ \ 1\ \ 1\ \ ...\ \ 1\ \ 0}_{\mathbf{x}_i}\right]^T$$

essentially, an unordered set of words without any knowledge as to the type of semantic resource for appraisal. The bag-of-words sentiment classifier operates on a simple word occurrence representation format. In this representation format, one (a full-text text analysis software program) counts the frequency of occurrence of a single word (unigram) or word group (bigram, trigram ... n-gram). As before, the word by document matrix is comprised of n words w_1, w_2, ..., w_n in m documents d_1, d_2, ... d_m, where the weights indicate the total frequency of occurrence of term w_p in document d_q. One additional row is added to this matrix: the labels y_i which denote the orientation of the document as either positive ($+1$), or negative (-1) or the subject category. To train for category, the labels y_i are 1, 2 or 3 for Product, Process or People, respectively. A sample vector used in the support vector machine SVM$_{sentiment}$ is shown in Fig. 5.5.

To train the support vector machine, sample design text that has been labeled with the correct sentiment and category must be produced. This is called tagged data. We produced tagged training and validation data for the sentiment classifier for a study on sentiment classification in design text (Wang and Dong 2007). Two cohorts of three native English speakers with a background in a design-related discipline (e.g., engineering, architecture, and computer science) were tasked with reading and categorizing various design texts. The rating cohorts were trained for one hour before they started formal rating work. The main purpose of training was to assure the cohorts could identify the proper category for each piece of design text and its semantic orientation according to the context.

During coding, 2 of the 3 coders had to agree on the semantic meaning (category), semantic orientation (orientation), and the attribute of the orientation, that is, positive or negative. Working in two hour time blocks, the coders read various design texts, including formal design reports, reviews of designed works, reviews of designers, and transcripts of designers working together. Every two hours, the coders took a fatigue check test to assess their performance. The fatigue test consisted of six appraisals that I had already labeled. The appraisals were randomized so that the group would not receive two tests containing the same set of appraisals. One-third of the data set was cross-categorized by another cohort until the two cohorts could agree upon a reliable rating. The cohorts' work was continually monitored to ensure that they were correctly following the rules for coding the categories and orientation according to the framework for the language of appraisal in design.

We would ideally like to have the classifier perform at least as well as the human coders in categorizing the text and its semantic orientation. In the practice of human labeling of data for training computational linguistic classifiers, the guideline of more than five votes per paragraph is used a baseline for confirming 'correct' labeling of a text. This coding system satisfies this requirement and is reliable because at least five people, four people from each cohort and one of the myself, agreed on the rating.

The performance of the coders with respect to accuracy of categorization (using the fatigue check test) of content was between 75% and 80% and, at times, dipped below 65%. They were at times inconsistent, which is a measure of their fatigue. In coding the clause "I was reeeally fussy with it" which should have been categorized in the People category because fussy is a state of mind, the coders categorized this clause as being about Process 15 of 18 times and as being about People 3 of 18 times. Their performance associated with semantic orientation was much better, with better than 90% accuracy over most of the sessions. Using the previous example clause again, they correctly identified this clause as being negative, as fussiness (at least in Australia) is considered negative. Only once did the cohorts think that fussiness was a positive characteristic. Pang (2002) found that human-based classifiers were accurate only 58% to 64%; thus, the performance of our coders is consistent with other studies and is likely to be the 'best' that could be expected.

Using the bag-of-words representation, i.e., a word-by-document matrix, the supervised machine learning system was trained on the design text corpus, $SVM_{category}$ for category classification and $SVM_{sentiment}$ for semantic orientation classification. On the target data set, we were able to obtain nearly 88% accuracy in sentiment classification (Wang and Dong 2008). Accuracy of categorization of the text according to Product, Process and People were consistent: 87% for Product, 84% for Process, and nearly 90% for People (Wang and Dong 2008) as shown in Table 5.8.

From these results and results from other researchers, the semantic orientation of documents appears to be largely dependent upon the semantic dependencies of words within a corpus, at least from a statistical natural language processing point of view. In a study done with my PhD student, we found that embedding any additional knowledge about the semantic orientation of individual words in a clause does not appear to improve or degrade the performance of the sentiment classification (Wang and Dong 2008). This result parallels findings in information retrieval which has shown that a purely statistical natural language processing approach such as latent semantic indexing (LSI) generally outperforms more knowledge-rich approaches.

The bag-of-words approach has practical downsides. The bag-of-words method relies on a very high-dimension representation that hinges on training a system on

Table 5.8 Performance of $SVM_{category}$

Category	$SVM_{category}$ Accuracy
Product	87.05%
Process	84.52%
People	89.82%

a text domain which contains a high coverage of words that are likely to appear in the target corpora. The fewer words that the two corpora share, the less likely it is that the bag-of-words sentiment classifier would perform well. This is not altogether attractive since this makes such systems difficult to transport across linguistic domains or even within a linguistic domain but for which the training set does not have as many unique words as found in the target domain. Also, if such a system were deployed on a very large corpus, such as the corpus associated with the design of very large engineering systems such as aircraft, it is very possible that there will be millions of features (unigrams). This very high dimensionality reduces the computational efficiency of the machine learning system and introduces other implementation challenges. Finally, the more unique words which exist in the text, the more training cases are needed, which is from a practical standpoint difficult to obtain.

Summary

The view in this chapter is that design is actioned with desire and intent. Desire and intent are emotional companions to design. The language of appraisal in design relays the force of affect through the desire-intent dyad. The language of appraisal in design is both an assertion of what is to be ordered and the level of disorder in the aggregated and accumulated language. In short, the language of appraisal has emotions. Semantic resources within the language of design key affect, which in turn affects the Product, Process and People. The core appraisal regions in language lay in the semantic resources for appraisal; grammar provides the structural connectivity between the appraisal regions and the conceptual regions (aggregation and accumulation). Aggregation and accumulation provide a materiality for appraisal; the appraisal is a response to the material that is being discursively produced as well as party to the creation of the material.

The language of appraisal in design can be conceptualized as a design guidance system that serves motivational and regulatory functions. In its motivational function, positive appraisals can signal what is desired to be made. In its regulatory function, it can evoke the production of new concepts when what has been produced does not yet satisfy the desire-intent litmus. In opposition to the motivational function, it can constrain what can be produced.

I end this section, and these previous three chapters, by recasting the discussion of the performative operators in terms of individual operators, as if the operators were a part of a decomposable system which is the language of design. The meaning potential of the language of design is by definition both conceptual, in the sense of having a linguistic 'body' produced by aggregation and accumulation, and affective. An important aspect of this model is that the performative operators do not work in isolation, but instead are part of a system. As the language of design enacts design, its actions should not be thought of as implemented solely by an individual performative operator, but rather by the interaction of the operators.

To understand the cross-over impact, their connectivity pattern needs to be determined. The cross-over is likely to include both structural connectivity, through semantic links and grammatical structures, and functional connectivity, in which the strength and polarity of one of the performative operators may attenuate or amplify the effect of another operator. At some point, we must acknowledge that the operators are conjoint and contribute in symmetric and asymmetric ways in the enactment of design. The viewpoint that might be promulgated by the one operator/one function perspective that has been taken up to this point is that they are mutually exclusive. I do not believe that there is one major role for each performative operator. The elucidation of the structural and functional interactions of the performative operators will be an important area of future research.

The performative operators are perhaps best thought of as codes that have regulatory principles that invoke, integrate, and select meaning potentials from language. How the codes combine to define modalities of design practice is the subject of the next chapter.

6 The Language of Design and Its Politics

> *[Robert] Moses used this phrase, so innocent in appearance,*
> *as authorization to write into contracts ... on the design*
> *and relative desirability of construction projects ...*
>
> Robert Caro, *The Power Broker: Robert Moses*
> *and the Fall of New York*, p. 705

Design, Language, Codes

We have presented the view that design designs through words, that the language of design performatively enacts design. The language of design can explicitly inscribe the designed work itself. As the epigraph reminds us, there are instances in which the language of design is performative in the Austinian sense – it was reportedly sufficient for Robert Moses to name a designed work to create the designed work. In many other instances, however, the patterns of language structuring are means by which design is produced by the performative operators of the language of design. The performative aspects of the language of design realize designed works which the language of design can speak of. The performative aspects write the relations that discourse about a designed work and design practice must establish. The performative aspects speak of and produce designed works by framing them (aggregating), forming them (accumulating), and rationalizing them (appraisal). In seeking a more capacious account imbricating the representational, constructive and instrumental roles of the language of design, our analyses have highlighted that the language of design is more than the use of symbols to designate design. Its performative operators render design. In doing so, language is seen not only as a message carrier but implicated directly in producing the carried.

Beyond the analysis of language realizing design practice and the designed work in describing it, we ought to consider if they are the only messages transmitted and produced through the language of design. Is there another layer operating with (perhaps 'behind') design discourse that regulates the way that designing can be (should be) accounted for? Arguing against essentialism, the concept of performativity claims that design is realized only to the extent that the performances cite prior, legitimized norms. The formation of the designed work and design practice takes place within a design praxis which set established traditions and obligations for those engaging in design. I have explained this issue to my more

A. Dong, *The Language of Design*,
© Springer 2009

'information technology-minded' peers in the following way. The architecture of network communication abstracts away the application protocol (e.g., HTTP, FTP) from the network protocol (e.g., IP) from the hardware protocol (e.g., Ethernet). This abstraction allows us to use a single type of data 'stream' to carry messages packaged for different application protocols. And, to a certain extent, the lower levels in the abstraction, such as the stream, control what is carried, such as breaking up a message into packets of data or wrapping the data with checks to ensure their integrity at the receiving end. Is the language of design just a stream of words that carries a message? Is it politically neutral or is the stream itself regulating what can be carried and how it can be carried? My argument in this chapter will be that the language of design is not ideologically free, and that accounts of design through the language of design are not necessarily fully willful accounts.

The political effects of the way that design is communicated are a topic that has not gone un-noticed. As I stated in the Preface to this book, much of what is known about design by 'non-designers' is gained through language-based communication about design – primarily through the media such as magazines about design, advertising, and television. Peter Lloyd, an academic in the Open University in the UK, conducts studies into the representation of design in the media, particularly television. In one study (Lloyd 2002), he analyzed three television programs to illustrate how the programs were effective in creating images of design – that is, lay knowledge about the process of designing and what designers do (including usefully dispelling the 'myth-conception' of the 'star designer' in depicting the cooperation of designers and design stakeholders). These television programs, as with other forms of mass media, are recruited to re-construct and reaffirm specific images of design. While the scene of recognition of design is the television program itself, there may be prior frameworks, such as Lloyd's training as an industrial designer, structuring how viewers judge and see design. Certainly, after watching one of these programs, people may develop new frameworks for seeing design. What I think studies such as Lloyd's are trying to locate are the sociology of design, that is, the structure and behavior of designers' relations with others and how these social relations affect the practice of design. The ways that design is communicated through language and the visual field are party to the sociology of design.

The question that I am approaching in this chapter is how the language of design can be seen as both producing and reproducing the social relations that are the backdrop to the sociology of design and that precede the accounts of design. This chapter raises the question how it was that the language of design 'taught' designers to become oriented to a particular socio-technical or socio-cultural system of design. Having highlighted the coupling between the language of design and the enactment of design, our attention now turns to how the language of design registers ideological assumptions that may regulate design and the way that designing can be acceptably characterized. What codes exist within the language of design that determine the way that design can be acceptably described and practiced? My contention is that if designers are giving accounts of design through language, then that language, and the 'canon' from which that language derives, becomes "a structuring structure which is in a continuous process of reproducing itself, mediating its identity through market forces, and negating the social conditions of its production by covering the tracks of its arbitrary and subjective formations." (Steiner 1996, p. 217)

As we discussed in the chapters on aggregation and accumulation, the language of design is drawn from both the particular design situation and a broad set of voices. What I am asking for is a conceptual apparatus that uncovers the ontological evidence for codes which govern how designers come to be in a position to frame and organize their way of accounting for their design practice and of orientating themselves to a design profession in a way that makes their mode of designing acceptable and recognized. Of particular interest is whether we could apply the same performative operators, aggregation, accumulation, and appraisal, to register these codes. The goal of the analyses that I will present in this chapter is not to get at empirical evidence to stake claims about the sociology of design, or even the sociology of specific design professions. Rather, the aim is to consider how the 'structuring structure' of the language of design could expose how the sociology of design disciplines is being developed.

The Sociology of Design Education

Design education, regardless of the discipline, includes rigorous study in domain competency, processes, tools, and, sometimes, humanistic concerns. While there are often well-developed curricula in technical design communication, these tend to focus on presentation of design information. What is overlooked is that the presentation participates in the language of design not just as material for overhead slides but as a realization of its codes. The way that language is used in the presentation affirms a sociology of design education. Language is more critical to design education than mere technical communication.

To explain why language use in design is important to the sociology of design education, we need to revisit the ideas of the Russian school of psychology and namely the ideas of Vygotsky. Central to Vygotsky's theories on learning is his rejection of biological explanations of learning. That is, Vygotsky argues that if we were to remove culture, function, and situated activity from learning, then learning could be reduced to biology – biology is the explanation for learning. However, this could not be, as Vygotsky concluded from observations through his studies on infant learning without language and the infantile use of language. In contrast to Piagetian theories on infant learning as progressively staged through the first two years of human development, Vygotsky theorized that cognition needs to be studied as a practice which arises out of a socio-cultural system in which the child's environment, including its objects, tools and people, operate as agents for the development of thinking skills. Language use is relevant because the means by which social interactions are co-ordinated become the means by which the child not only acquires symbols but also by which metacognitive competences for goal setting, planning and revising strategies for learning are constructed. The evolution of human linguistic competence notwithstanding (Tallerman 2005), Vygotsky approached the question of the role of language in learning from the perspective of language as a tool of cultural practice. His approach contrasts with the view of language as a grammatical system or language as encoding thoughts, which would be the basis of a linguistic or a psychological investigation, respectively.

What Vygotsky was searching for was a way to unify the elements of historical human cognitive development. Vygotsky theorized that cultural learning occurs in a zone of proximal development. Quoting Vygotsky's definition, Wertsch writes that "Vygotsky defined the zone of proximal development as the distance between a child's 'actual developmental level as determined by independent problem solving' and the higher level of 'potential development as determined through problem solving under adult guidance or in collaboration with more capable peers'." (Wertsch 1985, pp. 67–68) Schön's classic account of Petra learning with the studio master Quist is a prototypical example of the zone of proximal development in action wherein the studio master Quist performs "competences he would like her to acquire", (Schön 1985, p. 33) competences she would presumably not be able to acquire in his absence. For Schön, the zone of proximal development situates architectural education. Schön writes that architectural studio masters "make systematic descriptions of their practice and coaching, and the knowledge and appreciations embedded in them" (Schön 1985, p. 7) for the students.

The formulation of this difference allowed Vygotsky to characterize the transitions between the four historical domains of human development toward "higher mental functioning". Michael Cole characterizes Vygotsky's aim as a "full theory" of culturally mediated behavior that accounts for the "interaction of processes occurring at all the levels of human life system: phylogeny, cultural history, ontogeny, and microgenesis." (1995, pp. 191–192)

1. phylogeny – the evolution of the human species
2. history – the development of cultural tools and sign systems
3. ontogeny – psychological development
4. microgenesis – moment-to-moment changes of understanding when performing a task

According to Vygotsky, these four domains come together during social interaction. Social interaction is both a vestibule and an affordance for the transference of social capital. It is through such social interaction that cultural contents are transformed into differences in individuated cognitive processes.

In summary, Vygotsky theorized that children engage in a form of cultural apprenticeship to develop competence. Through socialization, the social becomes mental. Vygotsky, Luria, and Bakhtin proposed that language is a principal form of cultural mediation through which social interactions and cognitive structures are organized. As a significant refinement of Vygotsky's ideas while still maintaining the essential claims of Vygotsky's work, James Wertsch points out that meaning of words is a tool of mediated action rather than representing mediated action. In his words, "While continuing to accept his claims about the importance of semiotic phenomena in human mental functioning, I have argued that word meaning (or any other semiotic unit for that matter) is a unit of semiotic mediation of mental functioning, not a unit of mental functioning itself. I have also considered, and rejected as incorrect, the possibility that individual mental functions (memory, thinking) could serve as units of analysis in Vygotsky's approach." (Wertsch 1985, p. 208) That is, language is neither simply a linguistic system for the encoding and transference of symbolic knowledge nor to be construed as merely the outcome of cognitive structures that produce a linguistic output. Language is one of the means by which interactions in a social environment are coordinated and understandings of the world are discursively and socially negotiated.

The implication of these ideas is that language is both an agent of design and a mediating means for design. The Vygotskian notion of individual cognitive processes as internalized transformations of socially developed patterns of interpersonal interactions parallels our original motivation for delving into this exploration of the language of design as a way to get at the vectors of power that regulate the 'I/we' who is doing the designing. Language informs designers how to become oriented to a design praxis through codes that determine the way that design can be acceptably described and practiced in a design profession such as architecture or industrial design or engineering design. The question is whether the performative operators that enact design also set up implicit assumptions for the basis of the transformation of cultural variables into the identity of designers and their disciplines. Does aggregation set up expectations for the composition of actions, accumulation determine the allowable types of connections of these actions, and appraisal determine the criteria for significance of these actions?

We have already seen this behavior in the dialogue between Quist and Petra. As Quist is teaching Petra designing and learning about designing, what signals are available in the dialogue that allow Petra to abstract from the 'language of designing' (as Schön defines it) the codes that precede the language? Recall that for Schön, "Drawing and talking are parallel ways of designing, and together make up what I call the language of designing. The language of designing is a language of doing architecture, a language game which Quist models for Petra, displaying for her competences he would like her to acquire." (Schön 1985, p. 33) What I have been arguing thus far is that the language of design is not just paralleling design, not just describing design action as with Schön's "elements of the language of designing" (1985, pp. 44–45). Instead, the language of design has a role in tacitly registering the recognizability of a certain, legitimized design praxis.

When Quist is coaching Petra what to do next and different directions in which to take the design of the siting of the elementary school, he is not simply communicating design ideas or reframing the design problem. He is becoming a site for the relay and reproduction of reflection-in-action as a recognizable and legitimate design practice, putting the unregistered (and henceforth tacit) codes in the language of design to work as a way of claiming and norming his praxis. Reflection-in-action becomes part of the repertoire for architectural competence. More importantly, the procedure for gaining this architectural competence is social because it occurs through language-based communication and interaction. It is presumably a competence that can only be learned in a social context.

While Quist's speaking about design is an aspect of the Schönian language of design in as much as his speaking parallels the logos of his actions, he is establishing what Karl Maton describes as a 'knower mode' (2000) pedagogy. A knower mode pedagogy is characterized by claims to knowledge based upon a specific object of study, such as a designed work, by a 'privileged' person, such as a studio master, rather than a set of specialized procedures that have been institutionally demarcated (such as statistically validated hypothesis testing in the sciences). Maton claims through his analysis of the field of cultural studies that, in a knower mode pedagogy:

> Based on the unique insight of the knower, claims to knowledge by actors within the intellectual field are legitimated by reference to the knower's subjective or intersubjective attributes and personal experiences (which serve as the basis for professional identity within the field). ... This unique knowledge is specialised to the privileged knower such that actors with different subjective characteristics are unable to make claims about this knowledge, and attempts to do so risk evoking censure and even expulsion from the field. The knower mode thus exhibits strong classification and strong framing of its social relation. (Maton 2000, pp. 156–157)

From the position of pedagogic discourse, the language of design carries messages that allow the 'readers' to abstract 'codes' that precede the message. These codes determine whether the message is either a statement about design or some other statement. Referring back to the network communication metaphor, these 'codes' are like a 'checksum' as they serve to validate the content of the message. It is this second 'role' of language in design that we are interested in exploring in the context of pedagogic discourse in design education. To do so, we need to recover these codes that have been implicitly inscribed into the language of design. That is, an analytic device is needed for an analysis of pedagogic discourse in design. The eminent sociologist of British education Basil Bernstein calls such a device a 'language of description':

> [A] language of description is a translation device whereby one language is transformed into another. ... A language of description constructs what is to count as an empirical referent, how such referents relate to each other to produce a specific text and translate these referential relations into theoretical objects or potential theoretical objects. ... A language of description, from this point of view, consists of rules for the unambiguous recognition of what is to count as a relevant empirical relation, and rules (realisation rules) for reading the manifest contingent enactments of those empirical relations. (Bernstein 2000, pp. 132–133)

Bernstein is calling for analytical devices that describe aspects of pedagogic discourse by interweaving both empirical and theoretical descriptions and, simultaneously, a means for traversing between these two. That is, Bernstein would consider empirical descriptions of design as a socio-cultural system as an insufficient language of description if such a description were not accompanied by a theoretical reason why design is a socio-cultural system without relying on a tautological definition. Further, Bernstein argues for a conceptual way to move back and forth between the empirical and theoretical descriptions so that neither

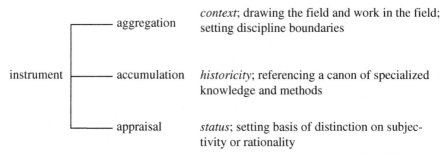

Fig. 6.1 Instrument to describe the transmission of design codes of practice through a sociology of design pedagogy

description remains impossibly estranged from each other. The language of design is for Bernstein a 'language of enactment' whereas the codes regulating this language offer a 'language of description'.

We have the basis for a language of description from the performative operators of the language of design. What is required now is to devise an instrument, using the same principles on the performativity of the language of design as before, to register these codes, thereby generating an empirical description. This device would then allow us to move back and forth between our theory of language and design and empirical descriptions of the sociology of design described by this device.

Figure 6.1 depicts an instrument for describing the pedagogic modality underlying design education in a specific discipline along the dimensions of aggregation, accumulation or appraisal. Rather than being viewed as producing a designed work, the performative operators here are regarded as producing a sociology of design and modes of behaving and orientating oneself within a design discipline. The instrument applies characteristic definitions for aggregation, accumulation and appraisal as in the previous chapters.

Table 6.1 depicts the corresponding coding scheme that can be applied to describe the underlying function of the language of design to regulate the social

Table 6.1 Coding scheme

Code	+	−
Aggregation	establishing and delineating disciplinary boundaries	opening of disciplinary boundaries; incorporating other disciplines
Accumulation	linking to established modes of practice and knowledge; design praxis linked to institutional or codified norms	practices not explicitly registered; specialized or specialist knowledge and procedures not demarcated
Appraisal	appealing to affect and subjectivity and designer's personal relation to design knowledge	appealing to technical rationality and empiricism

functioning of design (e.g., to establish a design discipline). Each code contains a set of characteristics for an analyst to locate representative empirical evidence within texts relating to the pedagogy of design. These determinants register how the message (e.g., pedagogy of design) that is carried in the language (e.g., peda-gogic discourse in design) is also regulating the sociology of design (education). Further, let us use the notation Ag+/Ag–, Ac+/Ac– and Ap+/Ap– to denote the modality of each code. The modality registers the direction of emphasis rather than the strength of emphasis.

Let us consider specific examples of texts on the teaching philosophy of disci-plines that teach design, comparing mechanical engineering, product design, and architecture. To sketch out how this instrument could indicate how a discipline of design registers the codes of norms, practices and currency, passages were taken from the Web sites of prominent schools. The semantics and grammatical struc-tures were 'read' with an attention to the determinants specified in the instrument. The italicized texts support the discussion that follows.

[MIT, Mechanical Engineering] Our educational mission is to prepare students for careers involving *technological innovation* [Ap–] and *leadership* [Ap+]. Our un-dergraduate educational program provides *a broad base* [Ag–] on which success-ful *careers in engineering* [Ag+] and *a number of other fields* [Ag–] can be founded, whereas the graduate program aims to *prepare specialists, professionals, and scholars in mechanical engineering* [Ac+]. The research mission of the De-partment – which is *to create knowledge, technologies* [Ac+] and *ideas* [Ac–] through *fundamental research and its application* [Ac+] – is closely intertwined with its educational mission.

The teaching and research programs in the Department are *organized* [Ag+] ac-cording to both *disciplinary* [Ag+] and *inter-disciplinary* [Ag–] themes. We cover *all of the core disciplinary* areas [Ag+] of mechanical engineering including *Dy-namics, Controls, Solid Mechanics, Materials, Fluid Mechanics, Thermodynam-ics, Transport and Design* [Ag+]. We have *strong interdisciplinary programs* [Ag–] in *Manufacturing, Energy, Bioengineering, Information, Nano/micro-Tech-nology and Ocean Engineering* [Ag+].

(http://www-me.mit.edu/GeneralInformation/Index.htm Retrieved 01-June-2006.)

[Parsons The New School for Design, Product Design] Through an *immersion in materials, processes, aesthetic consideration and proactive social engagement* [Ac+], Parsons Product Design Department *cultivates* [Ac+] the *technical skills* [Ap–] and *intellectual habits* [Ap+] essential to *imaginatively explore* [Ap+], *re-sponsibly integrate and synthesize* [Ap–] the *swiftly expanding roles* [Ag–] of a successful, professional product designer.

The Product Design Department prepares students for *a broad spectrum of profes-sional career directions* [Ag–]. In a three-year program, they learn to conceive *thoughtful* [Ap+] and *functional* [Ap–] domestic and consumer products, and make *intelligent, responsible* [Ap+] use of the latest technologies and materials.

While being introduced to *a variety of design methodologies and the history of product design* [Ac+], students *master the fundamentals of computers, machinery, and tools, as well as presentation and research techniques* [Ac+]. They are *taught the design processes* [Ac+] through which a product is conceived, developed, fabricated, and marketed, while *developing an awareness of New York City's professional and cultural resources* [Ap+]. Students are also *taught to investigate* [Ac+] the impact the products they design will have on the environment throughout their life cycles. They also *begin to grasp marketing strategies and theories of ethical practices* [Ag–].

(http://productdesign.parsons.edu/html/about_html/deptmission1.html Retrieved 01-June-2006.)

[Harvard Graduate School of Design, Architecture] *Central to the school's philosophy* [Ac+] is the *commitment to design excellence* [Ap+] that demands not only the *skillful manipulation of form* [Ac+], but also *inspiration from a broad body of knowledge* [Ac–]. Instruction and research *encompass design theory as well as visual studies, history, technology, and professional practice* [Ac+]. The GS's information infrastructure provides *a foundation for design exploration and communication* [Ac+], offering students *new ways to access design references, model buildings, and present ideas* [Ac+]. *Intelligence* [Ap+], *creativity* [Ap+], *sensitivity* [Ap+], and *a thorough knowledge of the arts and sciences* [Ac+] are essential to achieving *distinguished architecture* [Ap+]. The educational experience at the GSD is *enriched and broadened* [Ag–] by *close interaction* [Ag–] among the *departments of architecture, landscape architecture, and urban planning and design* [Ag+], as well as by *many other resources* [Ag–] at Harvard University and MIT. *Architects draw upon knowledge and experience gained from the past* [Ac+] while *adapting* [Ap+] to the changing needs of the modern world.

(http://www.gsd.harvard.edu/academic/arch/ Retrieved 01-June-2006.)

Table 6.2 provides a summary of the codes from the above analysis.

The description of mechanical engineering is focused on defining the disciplinary boundaries. The field is distinguished by "core disciplinary areas" which are derived from the accumulation of knowledge from the disciplinary fields cited and the accepted methods of "fundamental research and its application". However, the program includes "interdisciplinary programs" which are nonetheless listed as

Table 6.2 Frequency of codes from analysis of course descriptions of university design courses

Discipline	Ag+	Ag–	Ac+	Ac–	Ap+	Ap–
Mechanical Engineering	6	4	3	1	1	1
Product Design	–	3	6	–	5	3
Architecture	1	3	6	1	6	–

distinct disciplines in and of themselves. Conversely, architecture is strongly anchored to a historicity of praxis: "design theory as well as visual studies, history,
technology, and professional practice". In terms of personal disposition, students
will need to acquire (or already have?) the characteristics of "intelligence", "creativity" and "sensitivity". However, architecture is open to other fields and disciplines, and it is suggested by the description that many architecture students will
practice design in other disciplines such as product design. Product design, like
architecture, is strongly defining a core set of practices. Given that the discipline
of product design is often considered as situated somewhere between engineering
and 'design' in the sense of allied arts, it is also open to integrating multiple disciplines. Product designers have "swiftly expanding roles" and their exposure to
other fields such as marketing and ethics is intended to incorporate those disciplines into product design.

We could repeat the analysis for descriptions of the respective disciplines by
professional associations. To illustrate the similar results, let us first examine the
description of engineering by Engineers Australia.

[Engineers Australia] Professional Engineers apply *advanced skills* [Ac+] in the
*analysis and knowledge of science, engineering, technology, management and
social responsibility* [Ac+] to *problem solving and synthesis* [Ap–] in *new* [Ag–]
and *existing fields* [Ag+]. … Professional Engineers lead teams or work in them
and need to be *innovative and creative* [Ap+] to develop the best possible solutions. The engineer must frequently make *balanced judgements* [Ap–] between
design refinement, cost, risk and environmental impact. … *Top level mathematics,
physics and chemistry are highly recommended subjects* [Ac+].

(http://www.engineersaustralia.org.au/careers/occupational-categories/
occupational-categories_home.cfm Retrieved 01-June-2008.)

There is an emphasis on a specific body of knowledge and rational approaches.
Contrast their description with that provided by the Royal Institute of British Architects on the architecture profession.

[Royal Institute of British Architects] As professional experts *in the field of building design and construction* [Ag+], architects use their *unique creative skills*
[Ap+] to advise individuals, property owners and developers, community groups,
local authorities and commercial organisations on the design and construction of
new buildings, … Architects can be *extremely influential* [Ap+] as well as being
admired for their imagination and creative skills [Ap+].

(http://www.architecture.com/EducationAndCareers/BecomingAnArchitect/
BecomingAnArchitect.aspx Retrieved 01-June-2008)

I have chosen to code the clauses by a single code, but it is possible to code certain sentences with two codes, for example, "*analysis and knowledge of science,
engineering, technology, management* [Ac+, Ap–] and *social responsibility* [Ac+,
Ap+]". The choice of analytical depth does not change the main point: the codes

point to ideological sentiments about design which gives a discipline its distinguishing context. Summarizing these analyses, there is an in-built logic in these statements which contribute to the social practice of design. It is not difficult to see that there are differences in the way that these disciplines and their professions are described and that the codes offer a way to abstract away from them what is valued and emphasized in terms of specialized modes of design practice and design identity.

Considerations of the constitutive levels of aggregation, accumulation, and appraisal in the pedagogic discourse of design has significant implications in design education, particularly in the area of the gender and ethnic minority gaps in engineering design[1] and certain 'creative' design disciplines such as architecture. While much has been written and researched on the gender gap in engineering and computer science, one theme that continually arises is that discourse on engineering design tends to focus on the historicity of knowledge in the field rather than the role of the discipline in society, its broader implications, and how students can contribute their dispositions to the field. Such an observation is consistent with the frequency of Ac+ and Ag+ codes and few messages of subjectivity (Ap+) in the above analysis of the teaching philosophy of engineering.

In other design disciplines, there exist gaps between ethnic minorities in large multicultural societies, such as the rather small number of Aboriginal Australians who take up architecture as a profession. Conjectures abound that architectural design practice is seen as an elitist profession in which success (at least in terms of reputation) is strongly determined by subjective peer appraisal and cultivated dispositions[2]. Unless students believe that they have sufficient social capital or 'cultivated taste' in accord with the dominant (established) peer appraisers, then they may be dissuaded from the profession. Again, this is consistent with the frequency in which personal dispositions coded by Ap+ are mentioned in the teaching philosophy of architecture. In other elitist 'creative' design fields, such as the arts, these appraisers are often called the "gatekeepers", a term used in the Systems Model of Creativity by Mihaly Csikszentmihalyi (1996) to describe people who determine creative people by selecting them for inclusion within a field. We should note here that the bias for cultivated tastes can work in many directions, such as gender bias in cultural studies. While the reasons for disparities in the demographics of students who take up specific disciplines of design, and the arguments for them, are complex and tenacious, unarguably, the point is that different disciplines of design operate with a certain set of ideological sentiments how design practice should operate. Being able to explain the politics of disciplinarity to design students allows them to recognize the values of a discipline. With this

---

[1] The National Academies Press (www.nap.edu) has a rich set of freely available (online) publications dealing with the subject of gender equity and minorities in engineering and sciences.

[2] According to the US National Science Foundation Division of Science Resources Statistics (SRS) data on Women, Minorities, and Persons with Disabilities in Science and Engineering there were, in 2004, there were 207,000 architects and 1.8 million engineers in the US. 187,000 (90%) of the architects and 1.511 (84%) million engineers were white. This data, based on Bureau of Labor Statistics, Current Population Survey, 1994–2004, is not encouraging for diversity.

understanding, they can decide how a variation on repetitive practices within the discipline could give them an agency to realize an individual design identity.

While the values of the modality of the codes certainly vary within the disciplines, we should not discount the specialized criteria operating to define these disciplines and the expected codes and currency of conduct of design within them. Whether the authors of the Web site texts consciously intended to write about engineering or architecture as they did or whether the way that the pedagogy and practice has been defined by these institutions is the 'official' way that the disciplines operate is somewhat immaterial. It is not entirely immaterial since these institutions are producing activities within the area of the pedagogy of design and thus creating the pedagogic modalities. It is not immaterial since Engineers Australia is a professional body representing the discipline as it is practiced in Australia. Even if you disagree with me that what the values of the modalities are, that is not the point of the analysis. As Bernstein has argued, the visibility of what the transmitter (e.g., educational institution, media) intends for the student to acquire may not be visible. It may only be tacitly stated. The aim of the performative operators as codes to analyze pedagogic discourse in design is to provide explicit and consistent descriptions of the modality of the pedagogy in these disciplines.

The concern here is to sketch out a way to expose the underlying structuring principles of accounts of design professions as provided through discourse (in texts) about design education to illustrate how the theory of the performativity of the language of design can also be used to expose the ideologies behind 'design is' statements by design-related professions. The analyses here are neither intended to propose a conclusive statement about the aforementioned design disciplines nor to define how design is practiced by these disciplines. Rather, my intent is to present a style of analysis. Conceiving of design in terms of the performative operators of the language of design may clarify debates about the sociology of design that appear to differ at the level of individual, socio-technical, or socio-cultural but are possibly recurrent forms of similar principles but with different emphases on each principle.

According to Bernstein (2000), any pedagogic discourse, whether it occurs at the societal, institutional, or classroom level, is about a regulated function, namely the function of regulating the formation of a specific discourse of knowledge about and within a specific field. Bernstein proposes three types of rules of pedagogic discourse which regulate the formation of specific discourses: *evaluative* rules which perform the role of managing continuous evaluation of the discourse, *recontextualizing* rules which produce principles by which discourses can be brought into a special relationship with each other for the purposes of their selective transmission and acquisition, and *distributive* rules that regulate the relationships between power, social groups, forms of practice, and who may transmit knowledge and under what circumstances. Bernstein claims that the pedagogic discourse that takes place in a classroom will likely reproduce the codes that regulate the discourse within a field. That is, within an instructional discourse which aims to transmit specialized skills for a field, there will operate a regulative discourse which creates order (how knowledge is organized within a field and its modes of realization), relations (the extent to which the knowledge is kept apart (symbolic isolation) from other fields) and identity (how knowledge and knowledge possessors are recognized and labeled as a consequence of participating in a field).

We might question then whether the pedagogic discourse that we analyzed above, in producing an identity for the discipline, also operates at the level of conversation between designers in the same discipline? Do designers working in industrial design re-produce their discipline's order, relations and identity and does it differ from designers in architectural design?

Let's examine discourses between two sets of designers, an industrial design team comprised of product designers and engineers taken from the Delft Protocols Workshop, and the classic dialogue between Quist the studio master and Petra the student by Schön. Again, we will use the same codes as before, but this time looking for micro-level evidence of the reproduction of the pedagogic discourse that takes place at the institutional level in its production of the order, relations and identity of a discipline.

First, let us start with the Delft backpack design team. I will again use the same code names and the spirit of the codes from the scheme of Table 6.1. To relate the coding to the production of design content, I will, in this analysis, pay attention to

Table 6.3 Backpack design team discussion (Cross et al. 1996)

Speaker	Segment	Statement
Kerry	(t 1)	What do we need? I guess **we should look at their existing prototype**[Ac+], huh?
John	(t 2)	Yeah, em, let me think **we could also just sort of like try to quantify the problem** [Ag–] because what's your understanding of the problem first of all?
John	(t 3)	They uh **they want a combination market product this backpack mountain bike product** [Ac+] and they've made a prototype and it hasn't been they're not pleased with it so far and the users' tests have some in in fact **it would be nice if we could see those users' tests to em see what the shortcomings were** [Ac+]
John	(t 4)	It it sounds to me that **what they're looking for is not they're kinda looking for a an interface a thing that will allow you to carry or fasten an existing backpack to an existing mountain bike** [Ac+]
Ivan	(t 5)	Yeah
John	(t 6)	**Is that how you guys interpret it?** [Ap+]
Kerry	(t 7)	Well also **they've got this em Batavus Buster** [Ac+] that em
Ivan	(t 8)	**Sorta pops on the bike?** [Ac+]
Kerry	(t 9)	**We can make it a special mountain bike so it could have the stuff required attached something to it** [Ag+]
John	(t 10)	OK
Moderator	(t 11)	I should point out that the bicycle which we have in the room is not the Batavus Buster but it is a typical mountain bike but the backpack which we have is actually the HiStar backpack which is uh is to be designed for it
John	(t 12)	OK

aggregation as setting the limits of design content [Ag+] or meta-processing about design content [Ag–], accumulation as about utilizing prior knowledge [Ac+] or introducing new knowledge [Ac–], and appraisal as either subjective [Ap+] or technical [Ap–].

At the start of their discussion (Table 6.3), the team begins by agreeing upon what it is that they should be designing. In discussing the design objectives and constraints on the problem, the group consistently refers to prior knowledge [Ac+], in segment (t 1) (the initial prototype made by the company), (t 3) the customer feedback, and (t 7) and (t 8) referring to the Batavus Buster. The Batavus Buster is an existing line of mountain bikes upon which they must base their design solution. In referring to prior knowledge through factual accounting, they justify the designed concept according to prior data (knowledge). After John engages in meta-processing about the design process in (t 2), he tries to get the group's understanding of the design problem. What is interesting is that in soliciting the group about their understanding of the problem, he is not asking for their personal dispositions but rather their factual knowledge and understanding of the design problem. John justifies his understanding of the problem based on factual data that also delimits the scope of their design work in segments (t 3) and (t 4) in which he repeats the key customer needs and the design brief. What is most striking is that when John asks for the group's appraisal of the design objective in (t 6), rather than providing an affirmation of the proposal, Kerry responds through a further elaboration of the requirements with reference to the Batavus Buster and Ivan by specifically by linguistically technicalizing about fastening the pack onto the frame. Kerry and Ivan continue this type of commenting on their understanding of the design problem by examining the Batavus Buster. There is very little in the way of discussion of empathy for the user of the backpack, or personal goals in the design of the backpack. In what other researchers have characterized as the design team naming and framing the design problem before moving forward (Valkenburg and Dorst 1998), the team uses grammatical features in which the actor of the clauses is either the client (e.g., "they want a combination market product") or the design work itself (e.g., "it could have the stuff required"). Here, then, in the conception of the design objectives and constraints, we see the group re-producing a normative engineering mode of design praxis which is to refer to prior solutions and empirical validation. In fact, the group is essentially following the prescription set by Pahl and Beitz (1999) in which the first step in design is to collect information about the requirements and constraints that will be embodied in the solution.[3] This empirical, fact-based approach to design, set forth in the pedagogic discourse about mechanical design and in the textbooks on engineering design are then reproduced in the forms of discourse.

In contrast, let us examine the portion of the dialogue between Quist and Petra (Table 6.4) in which Quist assists Petra in re-forming the direction and hence

---

[3] Further, Pahl and Beitz define a design methodology as "a concrete course of action for the design of technical systems that derives its knowledge from design science and cognitive psychology, and from practical experience in different domains." (Pahl and Beitz 1999, p. 10) This is a positively accumulating (Ac+) definition.

Table 6.4 Conversation between Quist and Petra (Schön 1983, pp. 82–85). Reprinted by permission of BASIC BOOKS, a member of Perseus Books Group

Speaker	Segment	Statement
Petra	1	**I am having trouble getting past this diagrammatic phase** – [Ap+] I've written down the problems on this list. **I've tried to butt the shape of the building into the contours of the land there** [Ag+] – but **the shape doesn't fit into the slope** [Ap–]. **I chose the site because it would relate to the field there** [Ap+] but the approach is here. So **I decided** [Ap+] **the gym must be here** – [Ag+] so **I have the layout like this** [Ag+].
Quist	2	What other big problems?
Petra	3	**I had six of these classroom units** [Ag+], but **they were too small in scale to do much with** [Ap–]. So **I changed them to this much more significant layout** [Ap+] (the L shapes). **It relates one to two, three to four, and five to six grades** [Ag+], which is more **what I wanted to do educationally anyway** [Ap+]. What **I have here is a space in here which is more of a home base** [Ag+]. **I'll have an outside/outside** [Ag+] **which can be used** [Ap–] **and an outside/inside** [Ag+] **which can be used** [Ap–] – **then that opens into your resource library/language thing** [Ag+].
Quist	4	This is to scale?
Petra	5	Yes.
Quist	6	Okay, say we have introduced scale. But in the new setup, what about north-south?
Petra	7	**This is the road coming in here** [Ag+], and **I figured** [Ap+] **the turning circle would be somewhere here** – [Ag+]
Quist	8	Now **this would allow you one private orientation from here** [Ag+] **and it would generate geometry in this direction** [Ag+]. **It would be a parallel** ... [Ag+]
Petra	9	**Yes, I'd thought of twenty feet** ... [Ag+]
Quist	10	**You should begin with a discipline, even if it is arbitrary** ... [Ag–] **The principle is that you work simultaneously from the unit and from the total and then go in cycles** ... [Ag–]

objectives of her design work. This is a rather complex dialogue between Quist and Petra in which the contours of the dialogue have more nuances than the reflection than Schön wrote about. Specifically, I would like to draw attention toward how Petra describes and then justifies her actions. In the statements where Petra names the design content, I coded them as Ag+. What is intriguing is that she rationalizes the basis of the design content with appraisals that relate to her personal stance to the design content, coded as Ap+. Almost after every instance of naming a reference or a design move, she appraised the work or action relative to her subjective preferences. Specifically, she uses language such as "I decided" and "what I wanted" much more commonly than referring to objective data about the current state of the design such as "the shape doesn't fit into the slope". When she states that the layout is "what I wanted to do educationally" she is displaying her empathy for the students. Her subjective method of appraisal contrasts with the way that the backpack team made references to objective data such as the feedback about the initial prototype in the customer surveys. Her interpersonal negotiation of her attitude about her designed work with Quist using subjective stances seems entirely consistent with the Harvard GSD description of practitioners of architecture requiring sensitivity. In not interrupting the way that she describes designing, Quist is tacitly affirming her account. He then displays what Schön called reflection at the end of this dialogue, which I have coded as Ag– to indicate meta-processing on the design content as a response to her strong linguistic technicalizing to describe her design content [Ag+].

Most importantly, in modeling reflection in the final segment, he is modeling the production of an identity, specifically the identity of an architect. The identity is a function of a way that designing should be accounted for if that identity is to be legitimate. Quist creates the identity of an architect through the following means. Quist models for her what counts as a legitimate display of accounting about architectural design. He is performing what Bernstein calls evaluative rules, rules that define what standards must be reached within a field to count as knowledge. In this case, segment 10 is not just a reflection but rather a statement that in order to attain recognition within the field, she must begin with "a discipline" to "work simultaneously from the unit and from the total". Quist affirms that architectural practice is done by having a vision and a way of working. In countermanding rationality with reflection-in-action, Quist is distinguishing architectural design practice from industrial design engineering. In performing the meta-processing, he is grounding Petra to an accepted modality of practice.

As this section has shown, the politics of the language of design is not contained in the content alone. Instead, its politics are formed by a way of making and realizing pedagogic relationships. The analyses of the macro-level discourse of institutional texts on design disciplines and design pedagogy and the micro-level discourse of design conversations illustrate a way to utilize the performative aspects of the language of design as codes to show how designer identities and their relations to the discipline are an outcome of these codes. As such, Bernstein's theory allows our analysis of the politics of the language of design to attend to the production of designer identities and relations in a way that seeing design communication as about design content alone cannot. We thus circle back to our original question: who is the 'I/we' who is doing the designing, who is speaking and writing about doing the designing? The politics of the language of design create, legitimize, and reproduce boundaries between disciplines of design and what constitutes the field. It is the same performative aspects of the language of design, which designers 'use' to realize design, which reciprocally set limitations on the way that designing can be described in order to be called design. I suggest that the issues raised by the question of who is the 'I/we' who is doing the designing could be analyzed through Bernstein's theory of pedagogic discourse. The performative aspects of the language of design expose the semantics within design discourses that set up modalities for knowledge production, transmission and acquisition.

In terms of producing disciplines of design, the performative operators constitute a set of operating principles behind the language that describe how design disciplines shape the identity of their disciplines. Through emphases on different aspects of these operators, the various design disciplines can assign significance and meaning to the principles. By practicing design as to value disciplinarity (i.e., Ag+), they create a design discipline that emphasizes disciplinarity. Owing to the performative nature of these processes in enacting design and the designed work, the above analyses suggest that the mode by which design is realized, and the disciplines realized, is regulated by the negotiation of identity and authority through the emphasis of one or more of these processes.

Various design disciplines may emphasize knowledge differentially depending upon how knowledge is valued and legitimized within the specific discipline. One dimension of Karl Maton's work on Legitimation Code Theory (LCT) (2004; 2007) claims that 'specialization' is what makes someone or something different, special and worthy of distinction. Through the specialization, the actors in the field determine what should be considered the dominant basis of achievement within the field. The orientation toward knowledge within a field may follow either an epistemic relation (ER) or a social relation (SR). This dimension of legitimation code is based on the premise that every practice, belief or knowledge claim is about or oriented towards something and by someone, and so sets up an epistemic relation to an object (ER) and a social relation to a subject (SR), respectively. Each relation may be more strongly (+) or weakly (–) emphasized in practices and beliefs. These two relative strengths of emphasis together give the LCT codes. Different fields may emphasize these relations to different degrees, and, as a result, these relations may be represented as being stronger or weaker within a continuum of strengths.

This means that knowledge can be seen as specialized by its epistemic relation, by its social relation, by both or neither, depending on how the field has negotiated what counts as legitimate knowledge in the field. As a result, LCT proposes four possible codes for how knowledge is specialized and legitimized in a field: *knowledge code* (ER+/SR–), *knower code* (ER–/SR+), *elite code* (ER+/SR+) and *relativist code* (ER–/SR–). The knowledge code emphasizes procedures; the possession of specialized knowledge, skills or procedures is emphasized as the basis of achievement whereas the dispositions of authors or actors are downplayed. In the knower code, the emphasis lies on personal characteristics of the knower. The personal characteristics could be natural (e.g. 'genius'), cultivated (such as an artistic gaze) or socially based (such as a specific gender, e.g. feminist theory, or sexuality; e.g. queer theory). The elite code emphasizes both the possession of specialist knowledge and the 'right kinds' of dispositions. In the relativist code, neither knowledge nor dispositions are necessarily required. In the language of LCT, we might expect to find architecture as an elite code discipline, engineering as a knowledge code discipline, fashion as a knower code discipline, and probably no design discipline as being dominantly relativist.

Using the concepts of LCT on claims to knowledge, we might ask then whether design practices across disciplines exhibit their discipline's varying emphases to an epistemic or social relation to knowledge, that is, appeal to affect and subjectivity (SR+) or technical rationality and empiricism (ER+). What specific practices within a discipline tend to link to established modes of practice (Ac+) rather than non-specialist procedures (Ac–)? I suggest that the performative operators of the language of design as operating principles of design along with the LCT codes' description of a discipline's orientation to knowledge define axes for the production of different practices within each discipline. Each discipline will have set up a prior emphasis on each operating principle and mode of knowledge legitimation to negotiate the authority of the discipline and what counts as legitimate practice within the discipline. The negotiation sets up what can be recognized as legitimate design practice and the expectations for what designers must do in order for their behavior to be ontologically described as designing.

The outcome of different practices emerging from common operating principles of design and how knowledge and knowers are perceived in a design discipline is depicted in Fig. 6.2. Let's take aggregation as the sample operator which, for the purposes of this example, is characterized by an emphasis on drawing together a set of ideas that could be framed into a design concept [Ag+]. The choice of emphasis along each performative operator then leads to a choice of emphasis on the legitimation of knowledge. At the end node, we would expect to find the space of possible choices (practices) for Ag+ along the routes ER+/SR+, ER−/SR+, ER+/SR−, and ER−/SR−. Different practices relate to different realizations of the choices (made by actors in various design disciplines) to conform (or not) to the principles of specialized design disciplines. Nonetheless, we are more likely to find practices such as the theory of inventive problem solving (TRIZ) (Altshuller 1999) being developed within and arising out of engineering design rather than fashion design, and that artistic gaze is more commonly cited in architectural design discourse than in structural design, given what these fields emphasize. I am not suggesting exclusivity; certainly there are soaring examples of artistry in structural engineering such as the Golden Gate Bridge in San Francisco, Santiago Calatrava's Oriente Station in Lisbon, and Frank Gehry's Guggenheim Museum in Bilbao. The point is that there is an emphasis on different techniques in these different disciplines and that what is commonly valued in engineering is specialized methods rather than having a vision and an identity, which is valued in architecture[4].

While all designers practice design in its broadest sense – the intentional production of a material work to satisfy functional needs – they also perform design in different ways. Such differences can be understood as reflecting the various values, beliefs, and mores held by a design discipline, or what Strickfaden has called the "culture medium" (Strickfaden et al. 2006). These values and beliefs function as structuring principles which generate and organize design practices; they are related to what Pierre Bourdieu defines as 'habitus' (1983) in that these values become internalized codes which equip the designer to operate successfully within the 'rules of the game' of a design discipline. How knowledge is put to use to practice design within a discipline is thus premised on what counts as knowledge and what counts as a recognizable design practice within the discipline. The operating principles of design set the foundation for explicitly foregrounding these criteria.

---

[4] Having attended both engineering design and architectural design conferences, I have observed strong differences in the style and content of presentations. To summarize my observations, I would suggest that displaying artistic gaze at an engineering design conference or, conversely, very high technical prowess at architectural design conferences is unlikely to win you friends.

Fig. 6.2 An operating principle of design and its realization according to varying emphases on claims to legitimacy of knowledge

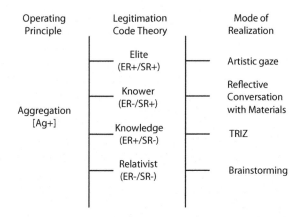

The possibility for different modes of realization to stem from a common set of operating principles provides a space for the intentionality of the designer to enter into the negotiation of the designer's identity within a discipline. Thus, we have a slightly more nuanced version of performativity in which the identity done by a designer relates to the discursive processes that constitute the mode of knowledge legitimation in a discipline of design. While a discipline may place emphasis on the knowledge code, for example, it is still possible for a designer to emphasize the knower code depending upon the designer's orientation to knowledge. While a discipline may emphasize strong disciplinary boundaries [Ag+], this does not preclude a mechanical engineer from working at or outside the disciplinary boundary, though this choice may result in the engineer finding it difficult to gain credibility within the discipline. In mapping between the operating principles of design and specific modes of realization of those principles, there can still be an element of intentional practice. The identity of the discipline does not foreclose the potential identity of the designer completely. Thus, while designers and designers' behavior will come about as "the resulting effects of a rule-bound discourse" within a system of power, our line of thinking is that the operating principles of design still give designers some agency in selecting the type of emphasis in realizing the operating principles. Variation along the codes allow the designer to be "A person [who] is not simply the actor who follows ideological scripts, but is also an agent who reads them in order to insert him/herself into them – or not." (Smith 1988, pp. xxxiv–xxxv). Even if the designer were performing (designing) in a theater (a discipline of design), the designer is nonetheless an actor with a personal history and who has some agency in choosing which theater to perform in. As such, the designer "has to be questioned as to its capacity for decisions, choices, interventions, and the like which are not specifically or solely determined by such categories as class or economics – however much they may be at the behest of ideology in general." (Smith 1988, p. 24) The performative operators as codes for organizing principles of the field of design along with the LCT codes give us the theoretical handles to impute the designer's choices for performing within an envelope of agency.

Non-cognitive Enactments of Design

Up to this point, the enactment of design by the performative operators is nonetheless achieved by a cognitive agent. We could call these cases cognitive enactments of design. We should be asking, however, whether design requires a cognitive agent. After all, computational models of design exist in which there is no cognitive agent other than the agent who assembled the algorithm. Simulated annealing and genetic algorithms are examples of such algorithms. Yet, even in these algorithms, we can see the performative operators at work. We find these operating principles for design operating in algorithms such as simulated annealing and genetic algorithms which have been adapted and adopted to enact design but are often not considered designing in and of themselves. Table 6.5 summarizes the relationships between the operating principles and simulated annealing and genetic algorithms.

Simulated annealing is a stochastic optimization algorithm modeled on the process of annealing in metallurgy. Often used for the optimization of a known design and for generating new designs, simulated annealing operates by formulating a design problem in an algebraic form. New design solutions are generated randomly through a stochastic, 'hill-climbing' algorithm. Design solutions are generated by modifying the value of a variable, and in doing so, enumerating the potential set of design solutions. One design solution is generated per iteration at a given temperature T. At each temperature, the quality of the solution is evaluated. If the quality of the solution is better than the prior solution, it is saved and modified in the next iteration. If it is not better than the prior solution, it might be saved depending upon the satisfaction of the Monte Carlo criterion at temperature T. The annealing schedule determines how quickly the system is 'cooled' and ultimately what 'imperfections' (non-optimal configurations) become 'frozen' into the solution. Thus, the stochastic design generation process looks like aggregation in the sense of finding areas in the design space which contain characteristics of the optimal solution and using those optimal characteristics to generate future solutions (accumulation). The annealing schedule determines the connectivity between the characteristics solidified into the designed work depending upon how quickly the system is allowed to cool. The Monte Carlo criterion is an appraisal of each solution generated.

Table 6.5 Operating principles for design operating in two computational design algorithms

Operating Principle	Simulated Annealing	Genetic Algorithms
Aggregation	Stochastic, 'hill-climbing' exploration of the search space	Genotype population production
Accumulation	Annealing schedule	Genetic operators of crossover and mutation
Appraisal	Monte Carlo criterion	Fitness function and roulette wheel selection

Genetic algorithms are a type of evolutionary algorithm based on the theory of evolution. Genetic algorithms evolve populations of design solutions. Each design solution (the phenotype) is encoded as a genetic string, typically in a binary string encoding, and together are known as chromosomes. Each bit in the chromosome is known as a gene. An encoded phenotype is called the genotype. In contrast to simulated annealing, genetic algorithms operate on multiple potential solutions simultaneously, or what is known as the population of candidate designs. An entire population of chromosomes is produced initially (typically randomly) and through various mechanisms of reproduction. At every generation, chromosomes can be modified through genetic operators such as crossover and mutation. Not all chromosomes 'replicate' to the next generation. A fitness function is applied to each chromosome and a technique known as roulette wheel selection is used to determine which chromosomes to use as parents for crossover and which chromosomes advance to the next generation. Modifications to the algorithm have been made to improve performance based on heuristics such as immigration to inject new genes into the population. Thus, the process of population production and heuristics such as immigration are akin to aggregation, systematic genetic operators that modify existing chromosomes are conceptually similar to accumulation, and fitness-based selection is a type of appraisal.

Speaking to colleagues who work in the area of genetics, I find that cell biology is a homologous domain to connect with these organizing principles from the language of design. Cell biologists use the terms aggregation and accumulation to describe cellular mechanisms associated with cell development. The issue of how non-cognitive models of design that enact design vis-à-vis the performative operators could be connected to cell biology is conjectural, but the conceptual views on cell biology seem consistent with the principles that have been described above. The mechanisms of human cell development are of course inextricably linked with life; with the widespread debate about stem cell research, the vocabulary of cell biology is infusing into the mainstream vernacular. In Chap. 3, we briefly discussed cellular aggregation. Links to our ideas on accumulation and appraisal in design also exist.

In normal development, when we are at the two cell stage just after fertilization, cells aggregate according to polarity and gravity comes into play. Polarity is defined in a biological context as the "persistent asymmetrical and ordered distribution of structures along an axis" (Cove et al. 1999). Polarity at the level of individual cells is central to the development of complex, multicellular organisms. The division of a polar cell generates non-equivalent daughter cells which eventually become differentiated cells. What is intriguing about cell development is that gravity is known to exert effects on the topographic relations of structures formed. The development of a polar axis often requires an input provided by a biological signal or by the physical environment such as light or gravity. The availability of space-based research where gravitational forces less than that on earth are possible allowed researchers to investigate the effect of microgravity on polarity in individual cells. Some studies in microgravity environments have shown that not all organisms require an input signal from the physical environment and that the effects of gravity may not be as significant in some organisms as it is for other organisms. Nonetheless, gravity clearly does have an effect and so gravity could be thought of as having an appraisal effect, directing the orientation of aggregated and accumulated cells. Hydrostatic pressure is also known to affect the morphology of yeast cells; different morphological structures in fungus arise due to environmental stimuli including heat and blue light.

Can we characterize the production of the multiplicity of cell structures as designing? I would tend to believe that cells are designing themselves and higher-order systems, with some inherent preprogramming, through the basic organizing principles of (cell) aggregation and accumulation (cell binding controlled by surface receptors, affinities for ligands on opposing species, adhesion processes and other biological variables) and the appraisal effect of external forces such as light, gravity, and pressure. Perhaps cells are enacting the organizing principles for design, principles which enable them to design new varieties of cells. Certainly, cells seem to be designing new forms of cancers for which the mechanisms of expression are not likely to be accountable for in terms of the mutation of genes alone. The functional pathways for gene expression effect different realizations depending upon a multitude of internal biological factors and external stimuli and continue to provide exciting challenges to biologists and geneticists unraveling their inner workings.

Many of the biological functions of cells are carried out by proteins. All of the information needed for the protein to carry out the function is contained in its amino-acid sequence which specifies the three-dimensional structure of the protein. Under a set of conditions, the proteins spontaneously fold into their native states which then enable them to carry out their biological functions. Here too we find a homologous domain to the organizing principles for the language of design.

Proteins are synthesized on ribosomes in cells based on the genetic program encoded in the cell's DNA. In order to produce the various chemicals that sustain life, the proteins undergo a process called folding. The folding of proteins into specific three-dimensional structures is a biological activity that has allowed living organisms to develop remarkable diversity and selectivity through underlying chemical processes. Proper protein folding and misfolding is linked with both healthy cell behavior and regulation of cellular growth and differentiation as well as disease. What is remarkable about protein folding is the role of the proteins themselves in promoting, regulating, modulating protein folding and disassembly, and their role in the degradation of proteins. Protein folding is dependent upon an abundance of certain chemicals (enzymes) to facilitate and catalyze molecular bonds. An unstructured protein can be formed from a chain that was newly synthesized by a ribosome or from a disordered aggregate. In this case, proteins are recruited into aggregations to provide the basis for folding into new structures (Dobson 2003). The existence (aggregation) of an abundance of these sets of chemicals is required for protein folding. Wright et al. (2005) propose in fact that the diversity of sequence identities between proteins plays a role in safeguarding proteins against misfolding and aggregation[5]. The folding topology then provides cavities which serve as active sites for binding; they are implicated in what we have been calling accumulation. The amino-acid sequence determines the "energy landscape" which determines when amino-acid sequences come into contact with each other, the bonding between the amide and carbonyl groups of the main chain (Dobson 2003). Finally, a series of molecular chaperones and folding catalysts (Gething and Sambrook 1992) recognize and modulate the state of folding. Critical to the prevention of disorganized prefibrillar aggregates which can harm cells, they neutralize such aggregates before they can cause disease. These chaperones behave in a manner similar to the performative operator of appraisal by promoting, inhibiting or reversing folding and assembly. They stabilize protein structures. They disentangle improperly folded proteins. They stabilize protein structures before assembly or translocation. The proteins, like words in the language of design, provide both the substrate for protein folding and the biological mechanisms of protein folding. All the information required for protein folding and unfolding is contained in the protein's polypeptide chain. Proteins, like words, realize their own narrative through self-assembly and dis-assembly, performatively producing specific protein structures.

I have taken some liberties in the description of protein folding using the conceptual apparatus of the language of design. Recent research in protein folding takes the view that the physics of protein folding are largely determined by the topology of a protein's native state (Baker 2000). The homology between the mechanisms of protein folding, or at least in the conceptual view of protein folding, and the organizing principles of design vis-à-vis the language of design, are rather surprising. While the prevailing view is that the energy landscape encoded in the amino-acid sequence evolved through natural selection, perhaps we could take the alternative view that the encoding has allowed proteins to design themselves. The physics and biochemical properties of proteins provide proteins with the capacity to enact design.

Our outlook suggests that design is an enactment of the operating principles of aggregation, accumulation and appraisal, derived from the performative operators of the language of design. Cognitive enactments are varied through the social structuring of knowledge. Non-cognitive enactments are varied through internal programming, such as in the genes constituting cells, enzymes and proteins associated with protein conformation, or the physics of crystal formation which follow the second law of thermodynamics. The functions of the genes involved in cell

---

[5] I should note that in this discussion on protein folding I use the term aggregation in a different way than biologists do when they discuss protein aggregation. Protein aggregation is (generally) an undesirable biological phenomenon arising from a population of partially unfolded proteins deposited in cells. Generally, the aggregation of misfolded proteins results in neurodegenerative diseases including Alzheimer's disease and Parkinson's disease, and type II diabetes. Protein aggregation is a competing process to protein folding. In my definition of aggregation, aggregation is taken to mean the collection of the base materials that provide a substrate for designing; in the domain of protein folding, my definition of aggregation concerns living cell environments containing the appropriate, diverse set of protein sequences.

development are not yet known nor are the mechanisms of external stimulus perception and their eventual effects clear. The complex and at times converging, interlocking, imbricating and cross-interactions of the performative aspects of the language of design and their expression as operating principles of design in cognitive and non-cognitive enactments of design make us question if design could be thought of as a set of abstract processes that can be enacted through a variety of systems and agents. As theorized by both Foucault and Deleuze, knowledge is afflicted by forces and systems which exceed knowers. Perhaps thinking of design as a capture of a flow of Deleuzian 'forces' in the form of the performative operators will allow us to make a new schemata for design that is capacious enough to include and inform those systems which enact design and yet delimit design from production systems; not every practice in the production of a reality is, after all, design.

Performance and Tension

In this final section, I would like to consider the dimension of the performative aspects that has to deal with their conditions of possibility. What are the conditions of possibility of the performative aspects in the language of design? My interest in this question is not to continue in an exercise in circular reasoning, wherein one might continue to ask for further underlying causes. Instead, my interest is in opening up a space for changes in power and control relations in design. In understanding what conditions of possibility propel or attenuate the language of design, we might then be in a position to appropriate these conditions in order to change the way that design is enacted.

My reaction is that we cannot look into the language itself for the answer. As we spoke about these performative aspects, we made a rather glib oversight about the possibility of existence of the performative aspects. While we have made arguments about their effect, we have as yet said nothing about how much power they need in order to produce their effects nor commented on the situational characteristics that set the stage for the formation of the performatives. There is a Bengali proverb that Amartya Sen cites in his book *Development as Freedom* that I think summarizes my point here. "Justice is like a cannon, and it need not be fired (as an old Bengali proverb puts it) to kill a mosquito." (Sen 1999, p. 254) Thus, the question that I would like to consider here is how design situations furnish the conditions for the emergence of the performatives. Do we always need the 'big gun' performatives in order to enact design?

One way that we can conceptualize this question is to think of the elements of performativity through a model. In this model, we need the following elements: a system that precedes the effect, a performative action, and the performance (the effect of the performative). Being a mechanical engineer by training, one example that comes to my mind is a string vibrating, particularly since the physics of vibrations will allow me to dramatize what sustains the performatives. Let us imagine that we have a string of constant length (fixed ends), gauge and material. This is our system. The tension, and I will return to this important concept of tension later, placed on the string is a condition that we can apply to the system. While the tension produces no manifest, it imposes a condition on the performance achievable by the string and any performative effect placed on the string. The string is set into motion by a performative action, being plucked; the performance (effect) is the harmonic vibration of the string. Under equivalent performative actions, the vibration in the string is different because of the varying tension in the string. The speed of the waves on the string depends on the string tension. Specifically, it is proportional to the square root of the tensions. Waves travel faster on a string with higher tension than another string and the frequency is therefore higher for a given wavelength. In musical instruments, strings under higher tension produce sounds different from strings under lower tension, all other aspects of the string being equivalent.

I propose that tension is a primary resource for the performative aspects of the language of design. Tension is the dimension that attenuates and amplifies the conditions of being of the performatives' effect. The performatives express the agency afforded by the tension. It is tension that propels the communicative reality.

Our analysis of the language of design has relocated the performance of design into the multiplicity of language. This is perhaps not so controversial. However, what opposes this model of design as discursively produced is not only the conflict between tradition and modernity and contemporaneity but also the intensification of coalitions of incompatible epistemological and political beliefs conjoined by the very process of designing. The latter point would suggest that there is no other model of design other than a discursive model, wherein various kinds of reasons and beliefs create a viscosity for the design situation. I am suggesting that it is in fact this viscosity, this tension, that propels the performative operators to subtend borders and muster the agency in order to produce an effect.

Thus, when the language of design is speaking, the language is not speaking from an individual, from a performer, to the participants, to the audience. The language of design is speaking about the relations between designer, participant, designed work, and praxis. The point then is not the model of performance but how the performance speaks of these modes of relations. What the performative operators display in their effect is the tension in which these relations hold.

I would like to briefly illustrate this concept of tension, performatives, and the language of design through two examples, one an electronic art installation and one a design cooperative. *Uzume* is an electronic art installation that uses a projection based virtual reality system known as a CAVE, developed at the Electronic Visualization Lab at the University of Chicago. Created by Roland Blach, Petra Gemeinböck and Nicolaj Kirisits, the production of the work is produced discursively between the work and the participant in the following way:

> This communication resembles the attempt to carry on a conversation with someone whose language one does not understand. Although the virtual environment reacts to the user's slightest movement, it does develop independently to a certain degree and challenges the visitor to try to figure out its peculiar linguistic code. (Blach et al. 2003)

Uzume's world is bound to the physical projection space of the CAVE and is based on the spatial representation of the temporal behavior of so-called "strange attractors". When the visitor moves about within the projection space, he/she crosses the attractors' parameter fields and thereby changes the respective state of their environment. Moreover, his/her presence triggers minute changes in *Uzume*'s medium, a force field in which both the user and the whirling structures are embedded.

The constitution of the performative effects of the participant's movements is made contingent on the viscosity of the force field. Whether or not the performative aspects are in the recesses of the dialogue between the participant and the art work is unknowable. What is manifest however is that the tension in the medium produces varying effects regardless of the regularity of the vocabulary of the dialogue.

Teddy Cruz is an architect whose work deals with the tension between the developed world and the developing world, with particular reference to the border between Tijuana, Mexico and San Diego, USA. Here, the tension that Cruz works with is not only the difference in economic and social strata in which the populations on either side of the border exist, but rather the attitudes towards being-in-the-world. Cruz writes of the generative possibilities of this tension in the following way:

> ... the growing tension between the various communities of San Diego and those of Tijuana have elicited a multitude of creative responses – new opportunities for sharing resources and infrastructure, for recycling at the most outlandish levels, and for normalizing local – not just international – relations between the cities of Tijuana and San Diego. (Cruz n.d.)

The above two examples suggest a politics of design(/er) identity and authorship, returning us to the original question posed at the start. Our response to this question is that the identity of the designer and the authorship of the designed work are intimately tied into the operating principles of design and the performative aspects of the language of design, aggregation, accumulation and appraisal. The designer, the designed work, design practice, and design praxis is not immune from the legitimizing codes that they produce. Kwame Anthony Appiah in his book *The Ethics of Identity* cited a quote that John Stuart Mill wrote regarding his book *On Liberty*. "When two persons have their thoughts and speculations completely in common; when all subjects of intellectual or moral interest are discussed between them in daily life ... when they set out from the same principles, and arrive at their conclusions by processes pursued jointly, it is of little consequence in respect to the question of originality, which of them holds the pen." (Appiah 2005, p. 26) Bearing in mind the social justice issues that Mill writes about, the issue becomes one of self-creation. Given the entanglement of the designed world with the creation of the matrix of conditions which form individual identities, the politics of the language of design is not only regulating itself; it is also expanding or diminishing the space of possibilities for political choices in realizing the designed world. Thus, it is not the I/we who is speaking about design that matters most and who it is and what the identity is; rather, it should be the embodied political choices for enacting design that matter. It is the set of choices (and what these choices are) which speak of the policy of design and of Bourdieu's formulation on the reproduction of social authority through embodied practices.

Michel Foucault wrote, "How is it that words, which in their primary essence are names and designations, and which are articulated just as representations itself is analyzed, can move irresistibly away from their original signification and acquire either a broader or more limited adjacent meaning?" (Foucault 1994, p. 109) This is the question that we have been wrestling within the context of design – how is it that words can become the designed work? How can words produce a content which is a designed work? What we find is that the language of design thus has several layers of features: communicative, performative, and political. As communicative linguistic events, language represents ideas in order to record and transmit them. Language use coordinates work, lexicalizes concepts, and represents the residue of cognitive processing. As a performative, the language of design produces the designed work and design practice. Through aggregation, accumulation, and appraisal, the language of design is constitutively involved in the production of design. Finally, as political acts, the language of design takes on the role of registering sociological and ideological conceptions.

In practice, design is characterized by social processes involving evaluation, reflection, and negotiations of shared meanings, leading to a shared understanding of both the design process and the design work. The design process progresses through stages including analysis, synthesis and evaluation. The effect of the design process is the design work. Each design project is a contestation of knowledge; the result of the contestation is the designed work. One could argue that all that is required is robust debate and negotiation to come to a shared understanding of the design work. Yet, this may be a meritocratic illusion for two reasons. Herbert Simon once wrote, "Seldom will the goals and constraints be satisfied by only a single, unique design; and seldom will it be feasible to examine all possible designs to decide which one is, in some sense, optimal." (Simon 1995, p. 246) Even if it were possible to negotiate to find the optimal design, the problem will be the negotiation process itself. The negotiation will necessarily presume that there is a shared ground as to what design is, what constitutes knowledge in design, and how the language should be realized to display competence in design. Yet, this may not be true. The question of what design is shapes who is viewed as having insight, who is entitled to participate in the negotiation, whose voice is more legitimate, and so on – whoever is able to claim the definition of 'design' can frame the negotiation. The consequence of discourses of design producing design according to the operating principles of design is that these negotiations will never be neutral because there is no single enactment of design. The choice of emphasis on the mode of knowledge legitimation defines different systems of discourse within design which results in different disciplines of design and different practices within each discipline. Each discipline negotiates the authority of the knowledge that has been generated and what counts as legitimate practice; each discipline, simply put, defines status in its own way. The negotiation sets up what can be recognized as legitimate design and sets up in advance the expectations for what designers must say and do in order for their behavior to be ontologically described as designing. As a consequence, we need to ask what kind (modality) of design is performed as the designer maps the operating principles of design to an individuated practice.

The language of design is an assemblage of performative aspects which systematically realize the objects which they describe. The question is whether a common system defines the emergence of these performative aspects or whether the performative aspects define a common system that regulates the emergence of meta-design phenomena. What is at stake in addressing this question, then, is not to examine what is said while designing but the generative conditions preceding the language of design and their effects *post-hoc*. If design is produced through rule-bound discourses, it will certainly be worthwhile to identify those rules.

7 Looking Back, Looking Forward

In the end I began to understand.
There is such a thing as absolute power over narrative.

Chinua Achebe, *Home and Exile*, p. 24

Looking Back

Consider for a moment a rather different way of conceiving of design. Chinua Achebe's quote in the epigraph continues as follows:

Those who secure this privilege for themselves can arrange stories about others pretty much where, and as, they like.

Narrative, for Achebe, is a form of political domination. It was a form that was used by English writers during the race for African colonization to depict Africans in ways which precluded Africans' ability to ethically present themselves in any form outside of the canon set forth in the narrative. Narratives promoted by English and European writers were used to justify the subjugation of Africans. Since Africans could not control this narrative, they were unable to control their own identities.

It is the narrative of design that this book has used to conceive of design, to give it an identity. One must also ask, as we did, where the narrative drew its power from and what order this narrative has produced. What codes within the narrative create processual, linguistic-oriented understandings of design? It is fairly clear, from my perspective, that the predominant narrative has historically originated from the view of design as the practice of professional designers, with slight discomfort in studying 'novice' designers except to differentiate them from the experts. In order to ascribe a set of practices as ontologically describable as design, a boundary has been drawn around a set of human practices. That boundary emanates from a discourse about design. It seems a simple formula enough. We find the underlying logic that since design is a practice done by humans, then how the human practices design should be a conceptual boundary. Starting from this premise, the design research community has located design practice in humans, with the occasional inspiration from biology (e.g., swarm intelligence) and post-structural theory (e.g., rhizomes). The definition of design and the set of

A. Dong, *The Language of Design*,
© Springer 2009

practices that constitute design nonetheless depend on the premise that design can be wholly located within the designer(s). Language has proven strategically effective in being the political form for coding a proper ontology of design.

I have proposed that we might be open to new possibilities for conceiving of design if we abandon the logic of design as a concept which is locatable in/from the body of the designer. I have asked whether it is time to reconsider whether design is primarily in the cognitive and social science ontology that grounds it. If language already metonymically refers to design, then perhaps the productive forces of language suggest us to invoke other imaginative resources about language in order to appreciate the vital dimensions of design itself.

It is these productive forces of language that form the logic of this book. Let us use the classic rule of transitivity to explain the logic. The rule of transitivity states that if you agree with me that A is equal to B, and you agree with me that B equals C, then you must agree with me that A equals C. Thus, we must start with the agreement that language (A) is performative, that is, reality-producing (B). Certainly, language 'is' lots of other 'things' but its reality-producing effects could not be argued away contemptuously. I doubt few would disagree that reality-producing (B) is *the* nature of design (C). Then language and design should be considered, perhaps at least conceptually, the same.

More to the point, I have been presenting that the language of design is performative, enacting and producing what it names. Performative is interpreted as the becoming of language into a designed work, a thing something other than itself. We have been working to locate the performative operators in the language of design. Empirical studies of language use in design by designers provided a 'validation' of sorts of those performative operators. I insist, however, that the aim has not solely been to conduct cognitive science, psychology or social science research computationally. The algorithms presented will likely find utility and resonance in these fields, or they may not. Be that as it may, the view that language is more than a byproduct or accompaniment of design is an attempt to provoke the bodily impediment to design studies.

This argument was articulated by a theory that there is a performatory character to the language of design. There are four key points to the theory of the performativity of the language of design:

1. The language of design enacts design through three performative operators: 1) aggregation – to frame lexicalized concepts that determine what will qualify as the material of design; 2) accumulation – to connect lexicalized concepts to achieve a transformation of words into a materiality; and 3) appraisal – to give affect to concepts as a means to co-construct and shape subjectivity.

2. The performative operators of the language of design locate themselves in the overall terminological patterns of design text and the interstices of their semantics and grammatical structures.

3. The 'felicity' of the performative operators map to 'successful' design outcomes.

4. The performative operators register how designers and design disciplines negotiate authority.

The theory takes as its premise that the structural units of any language of design consist of a set of symbols, a set of relations between the symbols, and features that key the expressiveness of symbols. The reality producing effect of language is itself an enactment of design, a praxis about materializing realities. Language enacts design through a set of information processing behaviors acting on the language of design – the performative operators of the language of design – aggregation, accumulation, and appraisal. These performative operators constitute a grid by which discourse on/about design produces design.

The interest has been in the abstract processes that the language of design performs rather than the linguistic system or cognitive structure producing the language. Our aim has been to examine the language of design by a careful study of how the language of design enacts design. The study grappled with the primal-dual question of how the language of design not only constitutes an account of designing and the designed work but also how the language of design harnesses and represents that which can be conversed and said, thereby becoming the act of designing and realizing the designed work.

The motivation to locate the abstract processes in design discourse that produce design was to uncover the operating principles of design, digging out shared 'rules' of design. We have done this because we have to be careful to acknowledge that the use of the language of design both acts upon, and is constrained by, the identity of the design discipline within which it is used. In turn, the field of design intertwines with language in a feedback circuit, affecting the structure of language as it enacts design.

The identity of designers is inscribed by the vectors of forces that come to inhabit their practice vis-à-vis words. When linked to theories on the social structuring of knowledge (e.g. LCT theory), each of the performative operators constitutes poles of designer behavior that links discourses about design to the modes of practices in design disciplines that arise from these poles. These poles of behavior provide an opportunity structure for the negotiation of a design identity by varying emphases on these poles. The various disciplines which recognize design as a unifying field and the multiplicity of design practices makes explaining the connections between design disciplines, practices and their codes an important endeavor.

The point here is that the semantics and grammatical forms of language use in design are not consequence free; they reflect the meaning potential of the designed work that is intended by the designer and the stakeholders. The meaning potential is not unregulated, though. It is encumbered by constitutive discourses. That these discourses are 'vocalized' or 'written' by designers makes the designers' language appear to be representing the nature of design. The discussion that this book opens is to think of design as a discourse-bound process that co-exists with intentional practices. The performative operators of the language of design are not intended to claim what design is, or how design should be practiced. Instead, we ask what ways design practice can multiply and arise from operating principles. It is the combination of these principles and intentional practices by which designers produce various forms of practice in non-arbitrary ways.

This exploration was undertaken through computational models of language processing[1]. Computation over the language of design is essential because the computation allowed us to interpret the 'performative operators' in the language of design as information processing actions that realize design. The use of computational linguistics is significant and not just a convenient methodological choice. When Claude Shannon published *A Mathematical Theory of Communication*, he showed that it was possible to model the generation of communication as a probabilistic system based on relatively simple rules on the statistical co-occurrence of letters in English words. The computational linguistics tools presented, latent semantic analysis, lexical chain analysis, and sentiment analysis, follow in Shannon's line of thinking. Latent semantic analysis claims that the statistical co-occurrence of words in discourse models the underlying concept and that meaning emerges from the statistical co-occurrence of words which frame the concept. The application of lexical chain analysis suggests that the statistical co-occurrence of semantic links in discourse reveals the way that ideas are connected by language and that concept formation is driven by the accumulation of lexicalized concepts. Finally, sentiment analysis enables us to extract a set of features from the language of design to recognize the way that the language of design 'feels'. The methodological conclusion is that it is possible to generate a computationally-derived view of how the language of design produces design based on the way that the language is assembled and functions. The worked examples of the computation of the language of design made it possible to obtain direct measurements of creative behaviors such as the social construction of knowledge, the formation of a concept, and the creative impulse.

Computing the linkages between the language of design and its reality-producing behaviors would not have been possible without first theorizing about how language produces realities through the theory of performativity. Table 7.1 summarizes the theorists from linguistic, political and cultural theory which informed the reworking of performativity to describe how language enacts design. The motivation to apply the theory of performativity has been the need to account for the extra-linguistic effects of linguistic praxis in design. The computational language

---

[1] There is a tendency to regard any computational system as entirely objective. As I have done in this book, the computational linguistic algorithms are related to their Other (critical theory) to remind readers of their subjective, philosophical links. Realizing the subjectivity of the computational linguistic algorithms does not mean that we should not dismiss the computational linguistic systems as having no practical value as instruments of empirical research. Rather, critical theory and computational science work in tandem: where critical theory gives us ways of thinking and problematizing, the computational linguistic based instruments lets us test our perceptions of reality, which we argued from critical theory, and so on.

Table 7.1 Performative operators, associated computational linguistic model and critical theorist(s)

Operator	Effect	Computational Language Model	Theorists
Aggregation	Gather material to form a frame for the design concept	Latent semantic analysis	Mikhail Bakhtin, Gaston Bachelard, Elizabeth Grosz
Accumulation	Build up the materiality of the design concept	Lexical chain analysis	J.L. Austin, Judith Butler, Eve Kosofsky Sedgwick
Appraisal	Negotiate subjectivity	Sentiment analysis	Jacques Lacan, Brian Massumi, Martha Nussbaum

models serve as both a description of how an ensemble of words operates to enact design and as a mirror-stage reflection for critical theory. The central claim that I am advancing is that it would not have been possible to explain why these computational language models had observed successes in modeling aspects of design without ascribing their technical performance to the capacity of an object (words) to enact what it represents or describes. The computational models synthesize the claim that words do what they say.

Looking Forward

This line of thinking has a second and perhaps more important implication. Language use in design is not neutral[2]. Design, in the words of Gui Bonsieppe, is part of an endeavor of social justice and not "after all, a tool for domination" (2006, p. 32). I could not agree more. Wanting design to be democratic and inclusive, though, does not mean that we can will design to be so. An official identity for design is always going to be constructed within design discourse, within the language of design. It may have an intrinsic regulatory and influencing function on what can be enacted. Beyond producing that which it names, it may legitimize what constitutes acceptable accounts of designing and 'recognizable' designed works. The ways in which the words in the language of design are realized impose limitations on what can be designed. We have shown this, literally, for concept formation using lexical chain analysis. In shaping the potential identity of a designed work, the language of design also shapes the behaviors and actions of the designers. What is said and written about designing and designed works reiterates the authority of what does and does not constitute design. Important as this is to design professions and formal design education, in situations ranging from the design of public infrastructure projects to community-oriented development projects,

---

[2] Luce Irigaray takes this line of thinking even further. She argues in her book *To speak is never neutral* that language is never neutral, that even an objective language such as science is sexed and neither neuter nor objective.

what can be positively achieved is directly correlated to the exercise of designer-like behavior by non-designers.

Basil Bernstein would argue that the language of design establishes codes or contains codes by which designers within a discipline share a common recognition rule which orients them to their specialty; so, it can exclude those who do not recognize these codes. Tacit codes within the language of design may inadvertently continue to affirm the tripartite of designers who design, the public (users) who provide input and consume the designed, and the policy-makers or capital holders who brace the authority to design to those schooled in design, the designers. It is the performative operators of the language of design that I have offered as candidate codes, not the exclusive codes, but ones that can be implicated with the enactment of design through language. These codes could provide a tool to analyze the formation of institutions, practices, and disciplines engaged in design, as illustrated in the previous chapter. It may then become possible to examine the relations between official pedagogic discourse about design in universities and formal curricula and the unofficial discourse in the public sphere.

There is one other important relation to Bernstein's theories. One concern that Basil Bernstein's œuvre dealt with is the transmission of pedagogic knowledge and its entwinement with economic and social class and the economic rationalization and capitalization of knowledge. Bernstein is careful to note though that his theory is intended to apply beyond the school context to any context in which relationships exist for the purpose of knowledge production-reproduction. One of the most socially progressive aspects of his theorizing deals with pedagogic rights. Bernstein proposes three pedagogic rights: 1) individual enhancement; 2) social, intellectual, cultural and personal inclusion; and 3) participation in a discourse and practice that must have outcomes (Bernstein 2000). What is remarkable about these rights is that they set the pre-conditions for the creation of new intellectual possibilities. If the transmission of knowledge is intended to do more than replicate an image of the knowledge reproduced onto a new subject, the transmission should refuse full coherence. Bernstein describes this incoherent transmission, the 'mystery of the subject', as that which allows for the potential for possibilities to be thought.

> By the ultimate mystery of the subject, I mean its potential for creating new realities. It is also the case, and this is important, that the ultimate mystery of the subject is not coherence, but incoherence: not order but disorder, not the known but the unknown. (Bernstein 2003, p. 240)

Design, ultimately, must be considered a resource that is valued for its potential to create new realities. The performative operators acting as codes within the field of design will, I hope, be useful in empirical and representative descriptions of official and unofficial practices of design. Rather than constraining design, yet not letting the concept of design become unnecessarily vague, my motivation is that the codes provide a more dynamic picture of the potential for and conditions of variation and change within the field.

Looking Far Ahead

Humans have been designing objects to suit a utilitarian purpose, to stimulate an emotive response, or to satisfy personal curiosity and inventiveness since at least our evolutionary Great Leap Forward. Interest in how we go about designing our world has paralleled our exuberance for designed artifacts.

We often wonder why a certain person or people appear to have an uncanny ability to design enthralling artifacts. We often think that they 'must be born with it'. To a certain extent, 'innate' talent appears to play a leading role for a number of noted designers. However, such arguments falter when we consider the number of people who are educated to become designers. Professional organizations and communities of practice convene to discuss methodologies for designing and to establish guidelines for the effective practice of their discipline. Books, journals, and academic curricula codify what is known about designing and how to teach the process. We also know that people designed objects before the time when designing was studied formally in universities. We know a lot about designing. And we know it's not just about talent.

Up until now, we have only addressed the issue of the connection between language and design as conceptual. Yet, surely, there might be a biological connection. It is said, though often disputed, that *only* humans have the capacity for both language and design. Granted, generally accepted definitions of language and design exclude any other species other than humans.

Yet, researchers in fields including genetics, neuroscience, and evolutionary linguistics often treat language and design as separate cognitive faculties, and more importantly, assume separate genetic links. Until only recently, neuroscientists generally accepted the existence of a specific language area in the brain, Broca's area. No similar design area in the brain has been identified or even postulated. Remarkably, the literature in these fields has very little to say in relation to any shared genetically-influenced capacity for language and design, although there is already evidence to suggest that design cognition requires the same cognitive capacities as language. There have been a number of computational studies using artificial life (Kirby 2002) to study language origins, few which mention design; yet, the creation of symbolic (language) and material (design) artifacts may likely be part of genetic traits supporting a set of semiotic abilities non-specific to language (Tomasello 1999). And, if genetically-influenced capacities for language exist, are these capacities shared with the capacity for design or are they different? Are there, for example, specific design competence genes? If so, do they express in a separate area of the brain?

The cognitive capacity for language (also referred to as linguistic competence) remains one of the most contested theories of human development. A key question is whether language is innate in the sense of a Universal Grammar (UG) shared across all humans, suggesting that the UG is encoded genetically, or whether language is purely an empirically evolved (human) phenomenon. In one camp are the so-called linguistic nativists, starting with Noam Chomsky and more recently

Stephen Pinker with his book *The language instinct*. In the other camp are the empiricists, most forcefully argued by Geoffrey Sampson and his response to Pinker's book specifically, and nativism broadly, in his book *The "language" instinct debate*. Terrence Deacon, in his book *The symbolic species*, argues for evolutionary-induced and conserved traits which compelled humans to develop language. Deacon claims that "symbolic reference itself is the only conceivable selection pressure ... Symbol use itself must have been the prime mover for the prefrontalization of the brain in hominid evolution." (Deacon 1997, p. 336)

It is probably wrong to believe that language evolved on its own as an independent cognitive capability; certainly, modern theories of the brain no longer support the concept that there is a single 'language' part (or module) of the brain. Philosophers such as Michael Devitt take a slightly more tempered approach, arguing only that "structure rules of a speaker's language are similar to the structure rules of her thought" (Devitt 2006, p. 275) but do not propose a strong link between the structure of language and the structure of thought which is the position of nativists and Pinker's concept of "mentalese".

The cognitive capacities enabling the generation of the semantics and syntax of language – orthography, phonology and semantics (Gitelman et al. 2005) – seem rather similar to the ones needed for design. The consequence of these capacities is humans' ability to represent objects (design) and abstract concepts (language) with arbitrary visual or material (design) and vocal (language) symbols and to act with reference to concepts not limited in time and space. Where vocal language realization requires dexterous manipulation of orofacial muscles and the larynx to articulate sounds, material design realization requires dexterous proprioceptive manipulation of muscles to handle the position or movement of a part of the body. Instead, it is more likely that a set of cognitive capabilities evolved simultaneously. Based on the evidence of tool forms which were not limited by mechanical constraints of the technique of manufacture, language is believed to have evolved at about the same time when tool making was unfolding (Davidson and Noble 1993). Arbib's Mirror System Hypothesis (2005) theorizes that the evolution of the language-ready brain proceeded at pace with the development of motor skills. For example, the ability to point to an object to refer to the object is likely to have evolved at the same time that the brain developed the capacity to refer to objects symbolically, which Arbib argued was essential for the formation of a protolanguage.

Other evidence of their link comes from anthropology. There is so much debate within anthropology over the record of human evolution, if you believe in evolution in the first place, that any discussion is risky. That being said, there is at least one piece of evidence that is decidedly intriguing. The theory comes from the work of David Lewis-Williams (2002). The world's oldest 'art' (about 77,000 years old) comes from the Blombos Cave on the southern coast of South Africa. What is intriguing about this art is that it was purely abstract. There were no iconic depictions of flora and fauna typically found in rock engravings and of the variety of cave drawings that comes to mind when we think of cave art. Researchers debate whether this important find suggests fully modern minds, language and symbolism at an unexpectedly early date. Lewis-Williams offers an alternative explanation.

Lewis-Williams argues that sensory deprivation in the caves, and possibly other factors, altered the states of consciousness of Upper Paleolithic *Homo sapiens*. If the brains of Upper Paleolithic *Homo sapiens* were fully modern, then they would have been able to engage in symbolic behavior in the form of decorative art (McBrearty and Brooks 2000). Given the sensory deprivation in the cave and the potential for the entry into a hallucinatory realm as a consequence of the deprivation, perhaps the abstract cave art is, in fact, a projection of mental imagery onto the surface of the cave. Lewis-Williams believes that what Upper Paleolithic *Homo sapiens* hallucinated was to them reality; the extraordinary event is that they then depicted what they considered reality onto the cave wall. The ability to experience altered states of consciousness is a psychobiological capacity of modern humans (Bourguignon 1979). The capacity to envision an altered state and to represent and to reify this reality would have been a significant precursor to activities that we now call design. After all, designing requires that designers envision a world that does not yet exist. This capacity to envision and reify symbolically would have been advantageous to Upper Paleolithic *Homo sapiens* over Neanderthals. Lewis-Williams contends that representational art of the variety found in the Blombos cave, along with other evidence of elaborate burial rituals, the use of red ochre over 285 kyr ago (McBrearty and Brooks 2000), and the use of marine pigment in symbolic behavior about 164 kyr ago (±12 kyr) (Marean et al. 2007), was used to mark off social groups. These social groups competed for resources for survival. Those who had the ability to form and keep social groups succeeded. Modern human behaviors including design, which McBrearty and Brooks term "technological innovativeness", arose gradually on an "as needed" basis. Modern human behavior, which includes abstract and symbolic thinking linked to art and design based on the archaeological evidence, either preceded language or appeared as a "package".

If language and design share intimate biological endowments, what genes are they? One gene of particular interest is the *FOXP2* (*forkhead box P2*) gene. This gene encodes an evolutionarily conserved transcription factor in the family of fork-head/winged-helix (FOX) transcription factors[3]. This transcription factor is associated with embryogenesis and is believed to be required for proper development of the regions of the brain associated with speech and language (Vargha-Khadem et al. 2005). The popular scientific literature called the *FOXP2* gene, associated with abnormal speech behavior the "language gene". The impairments and their neurodevelopmental genetic association were first recorded in three generations of the KE family (Lai et al. 2001). Affected members of the KE family, that is, those with the genetic mutation, have a severe impairment in the selection and sequencing of orofacial movements needed to articulate sounds. This is known as verbal dyspraxia. They are impaired in the realization of symbols and sounds to designate meaning (semantics and phonology). Further, they have severe impairments in language processing such as breaking up words into constituent sounds (phonemes) and grammar. It is thought that mutations in the *FOXP2* gene produce abnormal

---

[3] Transcription factors are regulatory proteins which bind to DNA and, in turn, control the transcription (or copying of the genetic information) of the gene either positively or negatively (Latchman 1997).

speech behavior evidenced neurologically by underactivation in Broca's area and other speech-related cortical regions, even under covert linguistic behavior in which verbalizations are thought but not said (Liégeois et al. 2003). A recent review of the literature concluded that the *FOXP2* genes are likely to be crucial to brain expression with respect to the development of speech and language capability (Vargha-Khadem et al. 2005). As one reads this review, I think you cannot be struck that the cognitive disorders associated with genetic neurodevelopment disorders affecting language competence would also impair design competence.

The *FOXP2* gene is not the only one likely linked to the linguistic competence. People with Williams syndrome suffer from a deletion of approximately 28 genes on chromosome 7q11.23 (Meyer-Lindenberg et al. 2006). Those with Williams syndrome present a very high sociability and empathy for others and eagerly engage in social interaction even with strangers. One peculiar behavior is an apparent heightened ability for language; people with Williams syndrome exhibit remarkable expressiveness and social communicative abilities given their other linguistic and cognitive impairments. Their apparently highly developed verbal skills are in fact a function of their diminished capacity to modulate the expressiveness of what is said as evidenced by under stimulation in the areas of the brain dealing with social judgment (Adolphs 2003). Sufferers of autistic spectrum disorders, and specifically autism, display impairment in communication and in particular a delay in language learning. They experience difficulties in the comprehension of relations between words (syntax and grammar) and symbols (orthography). Autistic disorders are linked to a series of genes (Freitag 2007). These diseases offer a unique window into their genetic influences on linguistic competence (Marcus and Rabagliati 2006). We might extrapolate that *FOXP2* and the forkhead/winged-helix family of transcription factors are a useful genetic starting point for investigating the biological interconnections between language and design.

It is even entirely plausible to conjecture that shared genetically-influenced cognitive capacities involved in language even influenced stylistic differences in the way cultures designed functionally similar objects. Dediu and Ladd's (2007) study on the distribution of tonal languages and the brain growth and development-related genes *ASPM* (*abnormal spindle-like microcephaly associated*) and *Microcephalin* is particularly relevant. Their study based upon linguistic population databases, statistical regression, and calculations of linguistic distance showed a nonspurious correlation between linguistic tone and the derived haplogroups of *ASPM* and *Microcephalin* (*MCPH1*), even after controlling for geography and history. Both of these genes are regulators of brain size and may have contributed to human brain evolution (Mekel-Bobrov et al. 2005). Computational simulations in linguistic diversity by David Nettle (1999) show that biological capacities coupled with social influences gave rise to the plethora of human languages.

It is therefore plausible that the genetic origins of language and design are interconnected. The debate over whether language and design are innate or whether they are learned centers around the question of how much specific information about language and design are hard-wired into the brain (by genes) and how much of the knowledge we acquire, that is, learn through social and cultural transmission.

Do we ascribe a strong genetic theory where genes trump culture and environment in the emergence of design competence or a weak genetic theory where culture more strongly influences the emergence of design competence? Regardless of which side is chosen, the pathway from the biology of design to the phenotype behavior of design is what we are actually after.

One significant difference between language and design is that no one is born with the ability to communicate with language. Language must be learned. Humans, like many other mammals, are born with some 'instinctive' ability to design, though we cannot (ethically) test this assertion. Certain 'designs' seem to be hard-wired into brains. We probably cannot even help but to design, at least, shelter and hand-held tools. It is remarkable that animals have an instinct to design. We can observe what they might be designing in their heads through objects they build:

- Chimpanzees born in captivity with limited opportunity to engage in or observe nest building activity in the wild nonetheless construct nests similar to those built by wild-born chimpanzees when both groups are provided with the same materials (Bernstein 1962).
- Gorillas, like humans, need support and protection from skin irritants when they rest. They build nests in trees and on the ground for this purpose, and there is little evidence of cultural variation in nests (Tutin et al. 1995).
- Beavers build dams to modify the landscape to increase its suitability for their occupation. They build dams by cutting wood and 'cementing' them with earth and mud (Gurnell 1998).

As Richard Dawkins stated in *The selfish gene*, we cannot flippantly wave away this instinctive behavior as 'instinct' (and important only to ethologists and zoologists) and therefore unimportant to humans. Rather, these mammals, particularly chimpanzees and gorillas which are highly genetically similar to modern humans, may contain vital genetic substrates that became candidates for exaptation during the evolution of cognitive capacities for language and design.

If we intersect the performative operators conceptually between language and design and biologically across neurodevelopmental disorders and their associated genes, a rather interesting picture starts to emerge. This intersection is described in Table 7.2.

We could hypothesize that design, which herein is viewed in its broadest sense as a mode of creative and inventive behavior, can be explored and understood through computational, agent-based systems in much the same manner as artificial life is used to understand language evolution and much in the same manner as we used computational linguistics to study the language of design. The emergence of design competence can be explained by general purpose learning systems and autonomous agents which inductively and adaptively learn from experience. This hypothesis thus brings us full circle. The phenotype behaviors expressed by the genes listed in Table 7.2 offer a biological basis to computationally model simple capacities to "bootstrap" agents that evolve toward design competence. Again, as with this book, we can turn to critical theory and computation to consider, now, the conceptual *and* biological interactions between design and language.

Table 7.2 Genetically-linked capacities for language and design based on language impairments and performative operators

Performative Operators	Language	Design	Disorder	Genetic Trait
Aggregation	Realization of symbols and sounds to designate meaning (phonology; semantics)	Realization of a set of ideas that can form the basis of a coherent design concept	Developmental disorders of speech and language	*FOXP2* mutation (Lai et al. 2001)
Accumulation	Comprehension of relations between words (syntax and grammar) and symbols (orthography)	Recontextualization of ideas to form design concept	Developmental disorders of speech and language	*FOXP2* mutation (Lai et al. 2001)
Appraisal	Modulation of expressiveness	Evaluation of design concept with respect to goals	Williams syndrome	deletion of ~28 genes on chromosome 7q11.23 (Meyer-Lindenberg et al. 2006)

References

Achebe C (2000) Home and Exile. Canongate Books, Ltd., Edinburgh

Ackerman JS (1986) The Architecture of Michelangelo. The University of Chicago Press, Chicago

Adolphs R (2003) Cognitive Neuroscience of Human Social Behaviour. Nature Reviews Neuroscience 4:165–178. doi:10.1038/nrn1056

Ahmed S, Wallace KM, Blessing LTM (2003) Understanding the differences between how novice and experienced designers approach design tasks. Research in Engineering Design 14:1–11. doi:10.1007/s00163-002-0023-z

Alexander C, Ishikawa S, Silverstein M (1977) A Pattern Language. Oxford University Press, New York

Altshuller G (1999) The Innovation Algorithm: TRIZ, Systematic Innovation and Technical Creativity. Technical Innovation Center, Inc., Worcester, MA

Appiah KA (2005) The ethics of identity. Princeton University Press, Princeton, NJ

Arbib M (2005) The Mirror System Hypothesis: how did protolanguage evolve? In: Tallerman M (ed) Language Origins: Perspectives on evolution. Oxford University Press, Oxford

Ausch R, Doane R, Perez L (2005) Interview with Elizabeth Grosz. Found Object. http://web.gc.cuny.edu/csctw/found_object/text/grosz.htm. Accessed 01 June 2008

Austin JL (1962) How to do things with words. Oxford University Press, Oxford

Bachelard G (1994) The poetics of space. Beacon Press, Boston, MA

Badke-Schaub P, Neumann A, Lauche K, Mohammed S (2007) Mental models in design teams: a valid approach to performance in design collaboration? CoDesign 3:5–20. doi:10.1080/15710880601170768

Baird F, Moore CJ, Jagodzinski AP (2000) An ethnographic study of engineering design teams at Rolls-Royce Aerospace. Design Studies 21:333–355. doi:10.1016/S0142-694X(00)00006-5

Baker D (2000) A surprising simplicity to protein folding. Nature 405:39–42. doi:10.1038/35011000

Ball LJ, Ormerod TC, Morley NJ (2004) Spontaneous analogising in engineering design: a comparative analysis of experts and novices. Design Studies 25:495–508. doi:10.1016/j.destud.2004.05.004

Benami O, Jin Y (2002) Creative Stimulation in Conceptual Design. 14th International Conference on Design Theory and Methodology. ASME, New York, DTM-34023

Bernstein B (2000) Pedagogy, symbolic control, and identity: theory, research, critique. Rowman & Littlefield, Oxford

Bernstein B (2003) Class, codes and control. Routledge, London

Bernstein IS (1962) Response to nesting materials of wild born and captive born chimpanzees. Animal Behaviour 10:1–6. doi:10.1016/0003-3472(62)90123-9

Berry MW (1992) Large Scale Singular Value Computations. International Journal of Super-computer Applications 6:13–49

Beyer H, Holtzblatt K (1999) Contextual design. interactions 6:32–42. doi:10.1145/291224.291229

Blach R, Gemeinböck P, Kirisits N (2003) Uzume, CAVE Application, Ars Electronica Linz GmbH, Linz. http://www.aec.at/en/archives/center_projekt_ausgabe.asp?iProjectID=12298. Accessed 01 June 2008

Boden MA (1988) Computer models of mind: Computational approaches in theoretical psychology. Cambridge University Press, Cambridge

Bonabeau E, Dorigo M, Theraulaz G (1999) Swarm Intelligence: From Natural to Artificial Systems. Oxford University Press, New York

Bonsiepe G (2006) Design and Democracy. Design Issues 22:27–34

Boujut J-F, Tiger H (2002) A socio-technical research method for analyzing and instrumenting the design activity. The Journal of Design Research. http://jdr.tudelft.nl/articles/issue2002.02/article5.html. Accessed 01 June 2006

Bourdieu P (1983) The Field of Cultural Production, or: The Economic World Reversed. Poetics 12:311–356

Bourguignon E (1979) Psychological anthropology: an introduction to human nature and cultural differences. Holt, Rinehart and Winston, New York

Bruce RF, Wiebe JM (1999) Recognizing subjectivity: a case study in manual tagging. Natural Language Engineering 5:187–205. doi:10.1017/S1351324999002181

Bucciarelli LL (1994) Designing engineers. MIT Press, Cambridge, MA

Buchanan R (1989) Declaration by Design: Rhetoric, Argument, and Demonstration in Design Practice. In: Margolin V (ed) Design discourse: history, theory, criticism. The University of Chicago Press, Chicago, IL

Burgdorf J, Panksepp J (2006) The neurobiology of positive affect. Neuroscience & Biobehavioral Reviews 30:173–187. doi:10.1016/j.neubiorev.2005.06.001

Butler J (1993) Bodies that matter: on the discursive limits of "sex". Routledge, New York

Butler J (1997) Excitable speech: a politics of the performative. Routledge, New York

Butler J (2005) Giving an Account of Oneself. Fordham University Press, New York

Cacioppo JT, Bernston GG (1999) The Affect System: Architecture and Operating Characteristics. Current Directions in Psychological Science 8:133–137. doi:10.1111/1467-8721.00031

Carroll JM (2006) Dimensions of Participation in Simon's Design. Design Issues 22:3–18. doi:10.1162/desi.2006.22.2.3

Chiu I, Shu LH (2008) Using language as related stimuli for concept generation. Artificial Intelligence for Engineering Design, Analysis and Manufacturing, 21:103–121. doi:10.10170S0890060407070175

Clancey WJ (1997) Situated cognition: on human knowledge and computer representations. Cambridge University Press, Cambridge

Cole M (1995) Socio-cultural-historical psychology: some general remarks and a proposal for a new kind of cultural-genetic methodology. In: Wertsch JV, del Río P, Alvarez A (eds) Sociocultural studies of mind. Cambridge University Press, Cambridge

Cortes C, Vapnik V (1995) Support-vector networks. Machine Learning 20:273–297. doi:10.1007/BF00994018

Cove DJ, Hope IA, Quatrano RS (1999) Polarity in Biological Systems. In: Russo VEA, Cove DJ, Edgar LG, Jaenisch R, Salamini F (eds) Development Genetics, Epigenetics and Environmental Regulation. Springer, Berlin

Coyne RD, Newton S, Sudweeks F (1993) A Connectionist View of Creative Design Reasoning. In: Gero JS, Maher ML (eds) Modeling creativity and knowledge-based creative design. Lawrence Erlbaum Associates, Hillsdale, NJ, USA

Cross N (1997) Creativity in design: analyzing and modeling the creative leap. Leonardo 30:311–331

Cross N (1999a) Design Research: A Disciplined Conversation. Design Issues 15:5–10

Cross N (1999b) Natural intelligence in design. Design Studies 20:25–39. doi:10.1016/S0142-694X(98)00026-X

Cross N (2006) Designerly Ways of Knowing. Springer-Verlag London Limited, London

Cross N (2007) Forty years of design research. Design Studies 28:1–4. doi:10.1016/j.destud.2006.11.004

Cross N, Christiaans H, Dorst K (eds) (1996) Analysing Design Activity. John Wiley & Sons Ltd, Chichester

Cruz T (n.d.) Border Postcard: Chronicles from the Edge, The American Institute of Architects, Washington, DC. http://www.aia.org/cod_lajolla_042404_teddycruz. Accessed 01 June 2008

Csikszentmihalyi M (1996) Creativity: flow and the psychology of discovery and invention. Harper Collins Publishers, New York

Cutkosky M, Engelmore R, Fikes R, Genesereth M, Gruber T, Mark W, Tenenbaum J, Weber J (1993) PACT: An Experiment in Integrating Concurrent Engineering Systems. Computer 26:28–37. doi:10.1109/2.179153

Davidson I, Noble W (1993) Tools and language in human evolution. In: Gibson KR, Ingold T (eds) Tools, Language and Cognition in Human Evolution. Cambridge University Press, Cambridge

Davidson RJ, Scherer KR, Goldsmith HH (eds) (2003) Handbook of affective sciences [electronic resource]. Oxford University Press, New York

de Vries B, Jessurun J, Segers N, Achten H (2005) Word graphs in architectural design. Artificial Intelligence for Engineering Design, Analysis and Manufacturing 19:277–288. doi:10.10170S0890060405050195

Deacon TW (1997) The symbolic species: the co-evolution of language and the brain. W.W. Norton, New York

Dediu D, Ladd DR (2007) Linguistic tone is related to the population frequency of the adaptive haplogroups of two brain size genes, ASPM and Microcephalin. Proceedings of the National Academy of Sciences 104:10944–10949. doi:10.1073/pnas.0610848104

Deerwester S, Dumais ST, Furnas GW, Landauer TK, Harshman R (1990) Indexing by Latent Semantic Analysis. Journal of the American Society for Information Science 41:391–407. doi:10.1002/(SICI)1097–4571(199009)41:6<391::AID-ASI1>3.0.CO;2–9

Deleuze G, Guattari F (1987) A thousand plateaus: capitalism and schizophrenia. University of Minnesota Press, Minneapolis

Devitt M (2006) Ignorance of Language. Oxford University Press, Oxford

Dhillon IS, Modha DS (2001) Concept Decompositions for Large Sparse Text Data Using Clustering. Machine Learning 42:143–175. doi:10.1023/A:1007612920971

Dobson CM (2003) Protein folding and misfolding. Nature 426:884–890. doi:10.1038/nature02261

Dong A (2004) Design as a socio-cultural cognitive system. In: Marjanović D (ed) Proceedings of the DESIGN 2004 8th International Design Conference. Faculty of Mechanical Engineering and Naval Architecture, University of Zagreb and The Design Society, Zagreb and Glasgow, 1467–1474

Dong A (2005) The latent semantic approach to studying design team communication. Design Studies 26:445–461. doi:10.1016/j.destud.2004.10.003

Dong A (2006) Concept formation as knowledge accumulation: a computational linguistics study. Artificial Intelligence for Engineering Design, Analysis and Manufacturing 20:35–53. doi:10.1017/S0890060406060033

Dong A, Hill A, Agogino AM (2004) A document analysis technique for characterizing design team performance. Journal of Mechanical Design 126:378–385

Dong A, Kleinsmann M, Valkenburg R (2007) Affect-in-Cognition through the Language of Appraisals. In: McDonnell J, Lloyd P (eds) Design Thinking Research Symposium 7. Central Saint Martins College of Art and Design, University of the Arts London, London, UK

Dong A, McInnes D, Davies KP (2005) Exploring the Relationship Between Lexical Behavior and Concept Formation in Design Conversations. 17th International Conference on Design Theory and Methodology. ASME Press, New York, DETC2005-84407

Dougherty D (1992) Interpretive Barriers to Successful Product Innovation In Large Firms. Organization Science 3:179–203

Eggins S (2004) An Introduction to Systemic Functional Linguistics. Continuum International Publishing Group, London

Ekman P, Friesen WV (1975) Unmasking the face: a guide to recognizing emotions from facial clues. Prentice-Hall, Englewood Cliffs, NJ

Ekman P, Friesen WV, O'Sullivan M, Chan A, Diacoyanni-Tarlatzis I, Heider K, Krause R, LeCompte WA, Pitcairn T, Ricci-Bitti PE, Scherer K, Tomita M, Tzavaras A (1987) Universals and Cultural Differences in the Judgments of Facial Expressions of Emotion. Journal of Personality & Social Psychology 53:712–717

Ellsworth PC, Scherer KR (2003) Appraisal Processes in Emotion. In: Davidson RJ, Scherer KR, Goldsmith HH (eds) Handbook of affective sciences [electronic resource]. Oxford University Press, New York

Ericsson KA, Simon HA (1993) Protocol analysis: verbal reports as data. MIT Press, Cambridge, MA

Eris Ö (2003) Manifestation of Divergent-Convergent Thinking in Question Asking and Decision Making Processes of Design Teams: A Performance Dimension. In: Lindemann U (ed) Human Behavior in Design. Springer-Verlag, London

Fellbaum C (1998) WordNet: an electronic lexical database. MIT Press, Cambridge

Fillmore CJ (1974) The future of semantics. Berkeley Studies in Syntax and Semantics. Department of Linguistics and Institute of Human Learning, University of California, Berkeley, Berkeley, CA

Florida R (2002) The rise of the creative class: and how it's transforming work, leisure, community and everyday life. Basic Books, New York

Foltz PW, Kintsch W, Landauer TK (1998) The Measurement of Textual Coherence with Latent Semantic Analysis. Discourse Processes 25:285–307

Forgas JP (ed) (2000) Feeling and Thinking: The Role of Affect in Social Cognition. Maison des Sciences de l'Homme and Cambridge University Press, Cambridge

Foucault M (1994) The order of things: an archaeology of the human sciences. Vintage Books, New York

Franks B, Rigby K (2005) Deception and mate selection: some implications for relevance and the evolution of language. In: Tallerman M (ed) Language origins: perspectives on evolution. Oxford University Press, Oxford

Freitag CM (2007) The genetics of autistic disorders and its clinical relevance: a review of the literature. Molecular Psychiatry 12:2–22. doi:10.1038/sj.mp.4001896

Frew W (2005) Desalination plant 'too important to debate'. Sydney Morning Herald.

Gagné CL (2000) Relation-Based Combinations Versus Property-Based Combinations: A Test of the CARIN Theory and the Dual-Process Theory of Conceptual Combination. Journal of Memory and Language 42:365–389. doi:10.1006/jmla.1999.2683

Garcia ACB, Carretti CE, Ferraz IN, Bentes C (2002) Sharing design perspectives through storytelling. Artificial Intelligence for Engineering Design, Analysis and Manufacturing 16:229–241. doi:10.10170S0890060402163086

Garcia ACB, Howard HC (1992) Acquiring design knowledge through design decision justification. Artificial Intelligence for Engineering Design, Analysis and Manufacturing 6:59–71

Gero JS (1990) Design prototypes: a knowledge representation schema for design. AI Magazine 11:26–36

Gero JS (1998) Concept formation in design. Knowledge-based systems 11:429–435. doi:10.1016/S0950-7051(98)00076-8

Gero JS, Maher ML (eds) (1993) Modelling creativity and knowledge-based creative design. Lawrence Erlbaum Associates, Hillsdale, NJ

Gething M-J, Sambrook J (1992) Protein folding in the cell. Nature 355:33–45. doi:10.1038/355033a0

Gilmore JH, Pine II BJ (1997) The Four Faces of Mass Customization. Harvard Business Review 75:91–101

Gitelman DR, Nobrea AC, Sontya S, Parrisha TB, Mesulam M-M (2005) Language network specializations: An analysis with parallel task designs and functional magnetic resonance imaging. NeuroImage 26:975–985

Goldschmidt G (1992) Criteria for Design Evaluation: A Process-Oriented Paradigm. In: Kalay Y (ed) Principles of Computer-Aided Design: Evaluating and Predicting Design Performance. John Wiley & Sons, Inc., New York

Goldschmidt G (1996) The Designer as a Team of One. In: Cross N, Christiaans H, Dorst K (eds) Analysing Design Activity. John Wiley & Sons Ltd, Chichester

Goldschmidt G, Tatsa D (2005) How good are good ideas? Correlates of design creativity. Design Studies 26:593–611. doi:10.1016/j.destud.2005.02.004

Grosz E (2001) Architecture from the outside: essays on virtual and real space. MIT Press, Cambridge, MA

Gurnell AM (1998) The hydrogeomorphological effects of beaver dam-building activity. Progress in Physical Geography 22:167–189. doi:10.1177/030913339802200202

Halliday MAK (2004) An introduction to functional grammar. Arnold, London

Halliday MAK, Hasan R (1976) Cohesion in English. Longman, London

Halliday MAK, Matthiessen CMIM (1999) Construing experience through meaning: a language-based approach to cognition. Cassell, London

Hayles NK (2002) Virtual Bodies and Flickering Signifiers. In: Mirzoeff N (ed) The Visual Culture Reader. Routledge, London

Hearst MA (2000) The debate on automated essay grading. IEEE intelligent systems & their applications 15:22–37. doi:10.1109/5254.889104

Hill A, Dong A, Agogino AM (2002) Towards computational tools for supporting the reflective team. In: Gero JS (ed) Artificial Intelligence in Design'02. Kluwer Academic Publishers, Dordrecht, 305–325

Hofmann T (2001) Unsupervised Learning by Probabilistic Latent Semantic Analysis. Machine Learning 42:177–196. doi:10.1023/A:1007617005950

Hutchins E (1995) Cognition in the Wild. MIT Press, Cambridge, MA

Iedema R, Feez S, White PRR (1994) Media Literacy. Disadvantaged Schools Program, NSW Department of School Education, Sydney

Innes JE, Booher DE (2004) Reframing public participation: strategies for the 21st century. Planning Theory & Practice 5:419–436. doi:10.1080/1464935042000293170

Irigaray L (1985) This Sex Which Is Not One. Cornell University Press, Ithaca, New York

Irigaray L (2002) To speak is never neutral. Continuum, London

Jacobsen K, Sigurjónsson J, Jakobsen Ø (1991) Formalized specification of functional requirements. Design Studies 12:221–224. doi:10.1016/0142-694X(91)90035-U

Jencks CA (1981) The language of post-modern architecture. Rizzoli, New York

Joachims T (2002) Learning to classify text using support vector machines. Kluwer Academic Publishers, Boston

Kan JWT, Bilda Z, Gero JS (2007) Comparing entropy measures of idea links in design protocols: Linkography entropy measurement and analysis of differently conditioned design sessions. Artificial Intelligence for Engineering, Design, and Manufacturing 21:367–377. doi:10.1017/S0890060407000339

Kirby S (2002) Natural Language From Artificial Life. Artificial Life 8:185–215. doi:10.1162/106454602320184248

Kleinsmann M, Valkenburg R (2008) Barriers and enablers for creating shared understanding in co-design projects. Design Studies 29:369–386. doi:10.1016/j.destud.2008.03.003

Klimoski R, Mohammed S (1994) Team mental model: construct or metaphor? Journal of Management 20:403–437

Kosofsky Sedgwick E (2003) Touching Feeling. Affect, Pedagogy, Performativity. Duke University Press, Durham, NC

Lacan J (1977a) The mirror stage as formative of the function of the I as revealed in psychoanalytic experience. Écrits: a selection. Tavistock Publications Ltd, London

Lacan J (1977b) The subversion of the subject and the dialectic of desire in the Freudian uncon-
    scious. Écrits: a selection. Tavistock Publications Ltd, London
Lai CSL, Fisher SE, Hurst JA, Vargha-Khadem F, Monaco AP (2001) A forkhead-domain gene
    is mutated in a severe speech and language disorder. Nature 413:519–523.
    doi:10.1038/35097076
Lakoff G, Johnson M (1980) Metaphors We Live By. University of Chicago Press, Chicago
Landauer TK (1999) Latent semantic analysis: A theory of the psychology of language and mind.
    Discourse Processes 27:303–310
Landauer TK, Dumais ST (1997) A Solution to Plato's Problem: The Latent Semantic Analysis
    Theory of Acquisition, Induction, and Representation of Knowledge. Psychological Review
    104:211–240
Landauer TK, Foltz PW, Laham D (1998) Introduction to Latent Semantic Analysis. Discourse
    Processes 25:259–284
Landry C (2000) The creative city: a toolkit for urban innovators Comedia, Near Stroud, UK
Langan-Fox J, Shirley DA (2003) The nature and measurement of intuition: Cognitive and be-
    havioral interests, personality, and experiences. Creativity Research Journal 15:207–222.
    doi:10.1207/S15326934CRJ152&3_11
Latchman DS (1997) Transcription factors: An overview. The International Journal of Biochem-
    istry & Cell Biology 29:1305–1312. doi:10.1016/S1357-2725(97)00085-X
Lawson B (1997) How Designers Think. Architectural Press, Oxford
Leclercq P, Heylighen A (2002) 5,8 Analogies per hour: A designer's view on analogical reason-
    ing. In: Gero JS (ed) 7th International Conference on Artificial Intelligence in Design. Klu-
    wer Academic Publishers, Dordrecht, The Netherlands, 285–303
Leech G, Rayson P, Wilson A (2001) Word Frequencies in Written and Spoken English based on
    the British National Corpus. Pearson Education Limited, Harlow, UK
Leonard D, Rayport JF (1997) Spark Innovation Through Empathic Design. Harvard Business
    Review 75:102–113
Lewis-Williams D (2002) The Mind in the Cave: Consciousness and the Origins of Art. Thames
    and Hudson, Ltd., London
Liégeois F, Baldeweg T, Connelly A, Gadian DG, Mishkin M, Vargha-Khadem F (2003) Lan-
    guage fMRI abnormalities associated with *FOXP2* gene mutation. Nature Neuroscience
    6:1230–1237. doi:10.1038/nn1138
Lloyd P (2000) Storytelling and the development of discourse in the engineering design process.
    Design Studies 21:357–373. doi:10.1016/S0142-694X(00)00007-7
Lloyd P (2002) Making a drama out of a process: how television represents designing. Design
    Studies 23:113–133. doi:10.1016/S0142-694X(01)00024-2
Love T (1999) Computerising Affective Design Computing. International Journal of Design
    Computing 2:http://www.arch.usyd.edu.au/kcdc/journal/vol2/love/cadcmain.htm
Lu SC-Y, Cai J (2001) A collaborative design process model in the sociotechnical engineering
    design framework. Artificial Intelligence for Engineering Design, Analysis and Manufactur-
    ing 15:3–20. doi:10.1017/S089006040115105X
Lucius RH, Kuhnert KW (1997) Using sociometry to predict team performance in the workplace.
    Journal of Psychology 131:21–32
Luck R (2003) Dialogue in participatory design. Design Studies 24:523–535.
    doi:10.1016/S0142-694X(03)00040-1
Luck R, McDonnell J (2006) Architect and user interaction: the spoken representation of form
    and functional meaning in early design conversations. Design Studies 27:141–166.
    doi:10.1016/j.destud.2005.09.001
Mabogunje A, Leifer L (1997) Noun Phrases as Surrogates for Measuring Early Phases of the
    Mechanical Design Process. Proceedings of the 9th International Conference on Design The-
    ory and Methodology. ASME Press, New York
Malecki EJ (2007) Cities and regions competing in the global economy: knowledge and local
    development policies. Environment and Planning C: Government and Policy 25:638–654.
    doi:10.1068/c0645

Marcus G, Rabagliati H (2006) What developmental disorders can tell us about the nature and origins of language. Nature Neuroscience 9:1226–1229. doi:10.1038/nn1766

Marean CW, Bar-Matthews M, Bernatchez J, Fisher E, Goldberg P, Herries AIR, Jacobs Z, Jerardino A, Karkanas P, Minichillo T, Nilssen PJ, Thompson E, Watts I, Williams HM (2007) Early human use of marine resources and pigment in South Africa during the Middle Pleistocene. Nature 449:905–908. doi:10.1038/nature06204

Martin JR (2000) Beyond Exchange: APPRAISAL Systems in English. In: Hunston S, Thompson G (eds) Evaluation in Text: Authorial Stance and the Construction of Discourse. Oxford University Press, Oxford

Martin JR, White PRR (2005) The Language of Evaluation: Appraisal in English. Palgrave Macmillan, New York

Massumi B (1995) The Autonomy of Affect. Cultural Critique No. 31, The Politics of Systems and Environments, Part II:83–109

Maton K (2000) Languages of Legitimation: the structuring significance for intellectual fields of strategic knowledge claims. British Journal of Sociology of Education 21:147–167. doi:10.1080/713655351

Maton K (2004) The wrong kind of knower: Education, expansion and the epistemic device. In: Muller J, Davies B, Morais A (eds) Reading Bernstein, Researching Bernstein. Routledge, London

Maton K, Muller J (2007) A sociology for the transmission of knowledges. In: Christie F, Martin JR (eds) Language, Knowledge and Pedagogy: Functional linguistic and sociological perspectives. Continuum, London

Matthews B (2007) Locating design phenomena: a methodological excursion. Design Studies 28:369–385. doi:10.1016/j.destud.2006.12.002

McBrearty S, Brooks AS (2000) The revolution that wasn't: a new interpretation of the origin of modern human behavior. Journal of Human Evolution 39:453–563. doi:10.1006/jhev.2000.0435

McComb SA, Green SG, Compton WD (1999) Project Goals, Team Performance, and Shared Understanding. Engineering Management Journal 11:7–12

McDonnell J (1997) Descriptive models for interpreting design. Design Studies 18:457–473. doi:10.1016/S0142-694X(97)00012-4

Medway P, Clark B (2003) Imagining the building: architectural design as semiotic construction. Design Studies 24:255–273. doi:10.1016/S0142-694X(02)00055-8

Mekel-Bobrov N, Gilbert SL, Evans PD, Vallender EJ, Anderson JR, Hudson RR, Tishkoff SA, Lahn BT (2005) Ongoing Adaptive Evolution of *ASPM*, a Brain Size Determinant in Homo sapiens. Science 309:1720–1722. doi:10.1126/science.1116815

Meyer-Lindenberg A, Mervis CB, Berman KF (2006) Neural mechanisms in Williams syndrome: a unique window to genetic influences on cognition and behavior. Nature 7:380–393. doi:10.1038/nrn1906

Minsky M (1975) Minsky's frame system theory. Proceedings of the 1975 workshop on Theoretical issues in natural language processing. Association for Computational Linguistics, Morristown, NJ, 104–116. doi:10.3115/980190.980222

Morris J, Hirst G (1991) Lexical Cohesion Computed by Thesaural Relations as an Indicator of the Structure of Text. Computational Linguistics 17:21–48

Mukhija V (2003) Squatters as developers?: slum redevelopment in Mumbai. Ashgate, Aldershot, Hampshire, England

Nettle D (1999) Linguistic Diversity. Oxford University Press, Oxford

New South Wales. Parliament. Legislative Council. General Purpose Standing Committee No. 5 (2006) A sustainable water supply for Sydney. Parliament of New South Wales, Legislative Council. General Purpose Standing Committee, Sydney

Newell A (1990) Unified theories of cognition. Harvard University Press, Cambridge, MA

Norman DA (2005) Human-centered design considered harmful. Interactions 12:14–17,19

NSW Department of Infrastructure, Planning and Natural Resources (2004) Meeting the challenges – Securing Sydney's water future. Parliament of NSW, NSW Department of Infrastructure, Planning and Natural Resources, NSW Government, Sydney

Nussbaum MC (2001) Upheavals of thought: the intelligence of emotions. Cambridge University Press, Cambridge

O'Toole M (1994) The language of displayed art. Leicester University Press, London

O'Toole M (2004) Opera Ludentes: the Sydney Opera House at work and play. In: O'Halloran KL (ed) Multimodal Discourse Analysis. Continuum, London

Olson GM, Olson JS, Carter MR, Storrøsten M (1992) Small Group Design Meetings: An Analysis of Collaboration. Human-Computer Interaction 7:347–374

Ortony A, Clore GL, Collins A (1988) The Cognitive Structure of Emotions. Cambridge University Press, Cambridge

Ortony A, Clore GL, Foss MA (1987) The referential structure of the affective lexicon. Cognitive Science 11:341–364

Pahl G, Beitz W (1999) Engineering design: A Systematic Approach. Springer, Berlin

Pang B, Lee L, Vaithyanathan S (2002) Thumbs up? Sentiment Classification using Machine Learning Techniques. 2002 Conference on Empirical Methods in Natural Language Processing (EMNLP). Association for Computational Linguistics, Morristown, NJ, 79–86. doi:10.3115/1118693.1118704

Parliament of New South Wales (2005) Infrastructure Implementation Corporation Bill 2005, Act No 89 of 2005, Legislative Assembly of New South Wales, Sydney, New South Wales, Australia

Pedgley O (2007) Capturing and analysing own design activity. Design Studies 28:463–483. doi:10.1016/j.destud.2007.02.004

Pennebaker JW, King LA (1999) Linguistic Styles: Language Use as an Individual Difference. Journal of Personality and Social Psychology 77:1296–1312. doi:10.1037/0022-3514.77.6.1296

Regli WC, Hu X, Atwood M, Sun W (2000) A Survey of Design Rationale Systems: Approaches, Representation, Capture and Retrieval. Engineering with Computers 16:209–235. doi:10.1007/PL00013715

Reynolds CW (1987) Flocks, Herds, and Schools: A Distributed Behavioral Model. ACM SIGGRAPH Computer Graphics 21:25–34. doi:10.1145/37402.37406

Rothery J, Stenglin M (1999) Interpreting literature: the role of appraisal. In: Unsworth L (ed) Researching language in schools and communities: functional linguistic perpectives. Cassell Academic, London

Rowe PG (1991) Design Thinking. MIT Press, Cambridge, MA

Sack W (2000) Conversation Map: An Interface for Very Large-Scale Conversations. Journal of Management Information Systems 17:73–92

Schank RC, Larry T (1969) A conceptual dependency parser for natural language. Proceedings of the 1969 conference on Computational linguistics. Association for Computational Linguistics, Morristown, NJ, 1–32. doi:10.3115/990403.990405

Schön DA (1983) The reflective practitioner: how professionals think in action. Basic Books, New York

Schön DA (1985) The Design Studio: An Exploration of its Traditions and Potentials. RIBA Publications Limited, London

Schön DA (1988) Designing: Rules, types and words. Design Studies 9:181–190. doi:10.1016/0142-694X(88)90047-6

Sen AK (1999) Development as Freedom. Oxford University Press, Oxford

Shah JJ, Smith SM, Vargas-Hernandez N (2005) Multi-Level Studies of Design Ideation Components. In: Gero JS, Lindemann U (eds) Human Behaviour in Design'05. Key Centre of Design Computing and Cognition, Sydney

Shaikh MAM, Prendinger H, Mitsuru I (2007) Assessing Sentiment of Text by Semantic Dependency and Contextual Valence Analysis. In: Paiva A, Prada R, Picard RW (eds) Affective Computing and Intelligent Interaction, vol. 4738/2007. Springer-Verlag Berlin Heidelberg, Berlin

Silber HG, McCoy KF (2002) Efficiently Computed Lexical Chains as an Intermediate Representation for Automatic Text Summarization. Computational Linguistics 28:487–496. doi:10.1162/089120102762671954

Sim SK, Duffy AHB (2004) Evolving a model of learning in design. Research in Engineering Design 15:40–61. doi:10.1007/s00163-003-0044-2

Simon HA (1995) Problem forming, problem finding, and problem solving in design. In: Collen A, Gasparski WW (eds) Design and systems: General applications of methodology, vol. 3. Transaction Publishers, New Brunswick, NJ

Simon HA (1996) Sciences of the artificial. MIT Press, Cambridge, MA

Smith P (1988) Discerning the subject. University of Minnesota Press, Minneapolis, MN

Solovyova I (2003) Conjecture and Emotion: An Investigation of the Relationship Between Design Thinking and Emotional Content. In: Cross N, Edmonds E (eds) Expertise in Design: Design Thinking Research Symposium 6. Creativity and Cognition Studios Press

Song S, Dong A, Agogino AM (2003) Time variation of "story telling" in engineering design teams. In: Folkeson A, Gralen K, Norell M, Sellgren U (eds) Research for practice: innovation in products, processes and organisations, Proceedings of the 14th International Conference on Engineering Design. The Design Society, Stockholm, Sweden

Sonnenwald DH (1996) Communication roles that support collaboration during the design process. Design Studies 17:277–301. doi:10.1016/0142-694X(96)00002-6

Steiner CB (1996) Can the canon burst? (Rethinking The Canon). The Art Bulletin 78:213–217

Stempfle J, Badke-Schaub P (2002) Thinking in design teams – an analysis of team communication. Design Studies 23:473–496. doi:10.1016/S0142-694X(02)00004-2

Strang G (1988) Linear algebra and its applications. Harcourt, Brace, Jovanovich, Publishers, San Diego

Strickfaden M, Heylighen A, Rodgers P, Neuckermans H (2006) Untangling the culture medium of student designers. CoDesign 2:97–107. doi:10.1080/15710880600647980

Stumpf SC, McDonnell JT (2002) Talking about team framing: using argumentation to analyse and support experiential learning in early design episodes. Design Studies 23:5–23. doi:10.1016/S0142-694X(01)00020-5

Suchman L (2000) Embodied Practices of Engineering Work. Mind, Culture and Activity 7:4–18. doi:10.1207/S15327884MCA0701&2_02

Sutherland Shire Council (2005) State Govt Again Fails to Consult Over Desalination Plant, Sutherland Shire Council, Sutherland. http://www.sutherland.nsw.gov.au/ssc/home.nsf/WebPages/B1AF74E8164C1315CA25708200319747?OpenDocument&Expand=2. Accessed 01 June 2008

Tallerman M (2005) Introduction: Language origins and evolutionary processes. In: Tallerman M (ed) Language origins: Perspectives on Evolution. Oxford University Press, Oxford

Tang JC, Leifer L (1988) A framework for understanding the workspace activity of design teams. Proceedings of the 1988 ACM conference on Computer-supported cooperative work. ACM Press, New York, 244–249. doi:10.1145/62266.62285

Tóibín C (2004) The master. Picador, Sydney

Tomasello M (1999) The cultural origins of human cognition. Harvard University Press, Cambridge

Tomkins SS (1962) Affect, imagery, consciousness. Springer Publishing Company, New York

Turney PD (2001) Thumbs up or thumbs down?: semantic orientation applied to unsupervised classification of reviews. ACL'02: Proceedings of the 40th Annual Meeting on Association for Computational Linguistics. Association for Computational Linguistics, Morristown, NJ, 417–424. doi:10.3115/1073083.1073153

Turney PD, Littman ML (2003) Measuring praise and criticism: Inference of semantic orientation from association. ACM Transactions on Information Systems 21:315–346.
doi:10.1145/944012.944013

Tutin CEG, Parnell RJ, White LJT, Fernandez M (1995) Nest building by lowland gorillas in the
Lopé Reserve, Gabon: Environmental influences and implications for censusing. International
Journal of Primatology 16:53–76. doi:10.1007/BF02700153

Valkenburg RC (1998) Shared understanding as a condition for team design. Automation in
Construction 7:111–121

Valkenburg RC, Dorst K (1998) The reflective practice of design teams. Design Studies 19:249–
271. doi:10.1016/S0142-694X(98)00011-8

Vargha-Khadem F, Gadian DG, Copp A, Mishkin M (2005) *FOXP2* and the neuroanatomy of
speech and language. Nature Reviews Neuroscience 6:131–138. doi:10.1038/nrn1605

Vernes SC, Nicod J, Elahi FM, Coventry JA, Kenny N, Coupe A-M, Bird LE, Davies KE, Fisher
SE (2006) Functional genetic analysis of mutations implicated in a human speech and language disorder. Human Molecular Genetics 15:3154–3167. doi:10.1093/hmg/ddl392

Visser W (2006) The cognitive artifacts of designing. Lawrence Erlbaum Associates, Mahwah, NJ

Vredenburg K, Isensee S, Righi C (2002) User-Centered Design: An Integrated Approach. Pren-
tice Hall PTR, Upper Saddle River, NJ

Wang J, Dong A (2007) How am I doing 2: Computing the language of appraisal in design.
Design for Society: Knowledge, innovation and sustainability, 16th International Conference
on Engineering Design (ICED'07). Ecole Centrale Paris, Paris

Wang X, Dong A (2008) A Case Study of Computing Appraisals in Design Text. In: Gero JS,
Goel A (eds) Design Computing and Cognition DCC'08. Springer, Dordrecht

Wertsch JV (1985) Vygotsky and the Social Formation of Mind. Harvard University Press,
Cambridge, MA

Wertsch JV (1991) Voices of the Mind: A Sociocultural Approach to Mediated Action. Harvard
University Press, Cambridge, MA

Whitelaw C, Garg N, Argamon S (2005) Using appraisal groups for sentiment analysis.
CIKM'05: Proceedings of the 14th ACM international conference on Information and knowledge management. ACM, New York, NY, 625–631. doi:10.1145/1099554.1099714

Wiebe JM, Bruce RF, O'Hara TP (1999) Development and use of a gold-standard data set for
subjectivity classifications. Proceedings of the 37th annual meeting of the Association for
Computational Linguistics on Computational Linguistics. Association for Computational
Linguistics, Morristown, NJ, 246–253. doi:10.3115/1034678.1034721

Wilde DJ, Berberet J (1995) A Jungian Theory for Constructing Creative Design Teams. 7th
International Conference on Design Theory and Methodology (DTM). ASME Press, New
York, 525–530

Wood III WH, Agogino AM (1996) Case-based conceptual design information server for concurrent engineering. Computer-Aided Design 28:361–369. doi:10.1016/0010-4485(95)00055-0

Wright CF, Teichmann SA, Clarke J, Dobson CM (2005) The importance of sequence diversity
in the aggregation and evolution of proteins. Nature 438:878–881. doi:10.1038/nature04195

Zeitz CM (1997) Some Concrete Advantages of Abstraction: How Experts' Representations
Facilitate Reasoning. In: Feltovich PJ, Ford KM, Hoffman RR (eds) Expertise in context.
American Association for Artificial Intelligence, Menlo Park

Index